D0990025

CERAMIC INNOVATIONS
IN THE 20TH CENTURY

Edited by John B. Wachtman Jr.
Director of Materials Science, retired, National Bureau of Standards,
Professor Emeritus, Rutgers University, and
Editor, The American Ceramic Society

Published by

The American Ceramic Society

735 Ceramic Place

Westerville, Ohio 43081

The American Ceramic Society
735 Ceramic Place
Westerville, Ohio 43081

Printed in the United States of America.

08 07 06 05 04 03 02 01 00 99 10 9 8 7 6 5 4 3 2 1

ISBN: 1-57498-093-9

Library of Congress Cataloging-in-Publication Data
A CIP record for this book is available from the Library of Congress.

The line drawing that appears on the cover is figure 1 of "Extrusion Method for Forming Thin-Walled Honeycomb Structures," U.S. Patent No. 3,790,654, assigned to Rodney D. Bagley for Corning Glass Works, Corning, New York. Used with permission.

For more information on ordering titles published by The American Ceramic Society or to request a publications catalog, please call (614) 794-5890 or visit <www.acers.org>.

Contents

PREFACE

This book was written as part of the Centennial celebration of The American Ceramic Society held in 1998. The work was done under the general advice and guidance of the Ceramic Innovations Panel whose membership was as follows:

Neil N. Ault
Robert J. Beals
Joseph E. Burke
Margaret L. Carney
David W. Johnson
William D. Kingery
Dale E. Niesz
Joseph A. Pask
William E. Payne
William R. Prindle
William H. Rhodes
Jack A. Rubin
Edwin Ruh
Richard M. Spriggs
John B. Wachtman, Chair

This panel reported to the Centennial Committee under the cochairmanship of David W. Johnson and Richard E. Tressler.

Several members of The American Ceramic Society staff provided assistance. Special thanks are due to Mary Cassells, Greg Geiger, Mark Glasper, Russell Jordan, Mark Mecklenborg, Pat Janeway, and Paul Holbrook, and Robin Chukes.

This compilation could not have been done without the work of the authors of the descriptions of specific ceramic innovations and advice from many others. Special thanks for comments and contributions also are due to Richard O. Hommel, James R. Johnson, Rudi Metselaar, David W. Richerson, and Rustum Roy. Joyce E. Merritt coordinated submissions from Corning Incorporated. Robert E. Moore coordinated submissions on refractories. Jean Francois Baumard and Thomas G. Reynolds provided help with contacts for many ceramic technical societies.

Part I

The Development of Modern Ceramic Technology

1

The Development of Modern Ceramic Technology

By John B. Wachtman

Introduction

Advances in technology constitute a key driver and enabler of the contemporary, highly competitive world economy. Ceramics play many essential rolls in today's technology and are likely to be even more important in highly specialized roles as technology continues to advance. Needs and opportunities for ceramics in the future are particularly evident in such areas as computing, communications, defense, medicine, pollution abatement, and transportation. A look at trends in technology in general in the recent past and the associated development of ceramic technology provides a perspective for thinking about the future.

This book presents a brief summary of major developments in ceramics in the past century and a set of short descriptions of important innovations in ceramics during that time. Many advances in ceramics were made in the 100-year period since the first meeting of The American Ceramic Society in Columbus, Ohio, in February 1899. The founding of the Society followed a decision taken at the National Brick Manufacturers Association meeting in Pittsburgh in February 1898 to form a group for scientific presentations. The Society's founding also was stimulated by the establishment of the first department of ceramics at The Ohio State University in 1894.

Progress in commercial technology sometimes results from the accumulated effect of many small advances too numerous to consider in a short work. Attention here is focused on advances that took place in large and notable steps. As part of the Society's centennial celebration a representative selection of the most important such innovations in ceramics has been identified. These innovations are listed in the next section with a brief description of each. "Ceramics" here is taken in the American sense to include glass. A ceramic innovation is an advance in ceramic science, technology, or art that was new in some way at the time it was introduced and that had a major beneficial impact in practice. These innovations sometimes had their effect in the direct use of ceramics and in other cases indirectly through the use of ceramics in processing or as components in products.

Several ways of thinking about the technical base that existed in 1899 and upon which the past century of ceramic innovations was built suggest themselves as follows. What was the status of ceramic science and technology in 1899? What was the status of other technical societies pertinent to ceramics? What was the status of the university departments dealing with fields related to ceramics? We discuss these questions below to provide background information in considering progress in ceramics in the 20th century.

Advances in ceramics in the 20th century have, of course, taken place in many parts of the world—especially in the United States, Europe, and the Far East. Our list of ceramic innovations and the authors of their descriptions reflects this international character. In discussing the technological background against which these ceramic innovations occurred we mention many international developments, but we give particular attention to conditions in the United States. This focus makes the subject more manageable. Moreover, worldwide conditions in technology generally parallel those in the United States (sometimes taking place earlier or later) so that the pattern of developments in the United States provides a useful broad perspective.

Trends in Ceramic Technology before 1899

In this section we discuss the events leading up to the state of ceramic technology and allied technologies in 1899 as a preamble to discussing developments in the next century. There is no sharp dividing line between the two centuries. The continuity of developments sometimes requires mention of events after 1899 in the present section to complete the line of thought. The American Ceramic Society was founded within a few years of the beginnings of many new industries that were to change the world and incidentally affect the field of ceramics greatly. We note that Alexander Graham Bell's first telephone call was made in 1876. The first motor car factory began operation in 1895, the same year that the first motion picture was shown. The first airplane flew in 1903. We give other examples later of industries that profoundly affected technology in general and ceramics in particular. Moreover, many technologies affected life style and public expectations of goods and services. These factors also had an impact on stimulating ceramic technology. The emerging field of ceramic science and engineering, and its companion The American Ceramic Society, stepped on stage at the turn of the century and was swept along in a century of unprecedented economic growth, ruinous wars, and profound social change. In considering the emergence of materials science and engineering Arden Bement has classified the development of technology in the United States into four periods: 1. The birth and development of engineering in the United States (1825–1900); 2. The evolution of a national research infrastructure (1900–1945); 3. The evolution of a national science policy (1945–1973); and 4. The intensification of global interdependency (1973-present).

The status of ceramic science and technology in 1899 was that of art and craft in the process of becoming partly a field of science and technology although art and craft remain important in ceramics to the present day. A brief discussion of the development of the art and craft of ceramics and the progressive development of an organized engineering technology with an associated science base also provides perspective on the situation in 1899 and the subsequent development of ceramics in the 20th century. A very brief discussion of major trends outside ceramics that have stimulated or enabled developments in ceramics also is necessary. Ceramic engineering is a core discipline for production of ceramics. Many ceramic engineers also have done research and some have been important innovators. However, ceramic technology is the product of people from a wide variety of backgrounds. A short list of important contributing disciplines (listed roughly in order of their development as disciplines) includes chemistry, physics, civil engineering, mechanical engineering, geology, metallurgy, chemical engineering, ceramic engineering, and materials science.

Ceramics have been made and used since before the beginning of written history. Low-fired ceramic figurines as old as 25,000 BC have been found in Czechoslovakia. Pottery used as containers was made by the

Jomon of Japan by 8000 BC. A large body of practical art existed in 1899 and was the basis for much of the manufacturing of ceramics at that time. By 1899 the beginnings of ceramic science also existed. In Germany, H. Seger in 1874 contributed the "rational analysis" of ceramic raw materials into clay, quartz, and feldspar with various accompanying rules as well as developing his pyrometric cones for better firing control. In the United States, Josiah Willard Gibbs enunciated the phase rule in 1878. Worldwide the traditional craft of pottery and tableware making was influenced in the late 19th century by better bodies and glazes as well as better understanding of the effect of firing conditions. The electrical industry required durable electrical insulators. Samuel Morse initiated communication by telegraph between Baltimore and Washington in 1844. Bell-shaped glass insulators for telegraph poles were in mass use by the 1850s. The rise of the electric power industry brought feldspathic porcelain insulators in many forms into use. The automobile developed in Germany. In 1885 Karl Benz built his first motor vehicle using an Otto-type engine with a spark plug including a ceramic insulator. The chemical industry required porcelain for containers and liners. Many chemical processes were conducted in ceramic containers or containers lined with ceramics (e.g., in towers for sulfuric acid production). The iron and steel industry was led by England and grew explosively in the 19th century. It was a pacing contributor to construction, transport, and machinery. Magnesite refractories were introduced for the Bessemer converter that inaugurated the age of steel in 1856. The Bessemer converter was followed soon by the open-hearth furnace. Fireclay (high-alumina–silica clay) refractories became the workhorse. Brick and tile for general construction have been used since prehistoric times, but many types now common have come into mass use since about 1850. Solid brick are too heavy for use throughout tall buildings and hollow brick came into general use as mechanical methods for making them were developed. The original kilns for firing brick were bottle kilns and round down draft kilns. The first continuous long-chamber kiln (tunnel kiln) was built in Denmark in 1839. Producer gas was used to fire a tunnel kiln in 1873. Machinery for the mixing of ceramic raw materials and forming of unfired ceramics was originally worked manually. In England, James Watt's steam engine was developed for steam-driven mills, breakers, and mixers for ceramic raw materials. Many advances in machinery for ceramic production including presses and auger machines were introduced in 1850–1870. As electric motors and internal combustion engines became available late in the 19th century they replaced steam in powering industrial equipment, including ceramic processing equipment.

The use of lime mortar predates written history, but the prehistoric mortars would not harden under water. The Romans found that a mixture of lime and volcanic ashes would develop strength under water. Today's cement industry goes back to the studies of John Smeaton in England in 1776 in connection with the building of the Eddystone lighthouse. He found that, by firing soft, clayey limestone, he obtained a material that would harden under water. The firing converted calcium carbonate at least partially to calcium oxide that would hydrate under water and so harden. A wide range of compositions of hydraulic cements is possible and a series of investigators added other raw materials. The invention of portland cement (distinguished by containing silica and alumina in addition to calcium oxide) is usually credited to Joseph Aspdin in England in 1824. Reinforced concrete was first made in 1868, and the rotary kiln for cement making was invented in 1873. The manufacture of portland cement in the United States began in 1875.

The long process of moving ceramics from a tradition-based craft to a science-based technology conducted under the direction of engineers was underway in the 1800s and has continued to the present day. The United States imported its scientists and engineers in its early years. Although the United States continues to this day to import many technically trained people, a very strong domestic network of universities and technical societies was gradually built up. The development of science and engineering in the 19th century was accompanied by the founding of specific departments in universities and technical societies dedicated to newly emerging fields of science and technology. Some highlights follow. The first field of engineering taught at college level in the United States was military engineering as taught at the United States Military Academy (West Point), which was founded in 1802. The building of the Erie Canal (1817–1825) stimulated civil engineering and led to the founding of Rensselaer Polytechnic Institute in 1824 and the teaching of civil engineering there. Technical departments in higher education developed extensively during the 19th century. Not the least of President Abraham Lincoln's achievements was his successful support for the

Morrill Land Grant Act of 1863. This Act provided support for colleges in "subjects relating to agricultural and the mechanic arts," which should be made "leading branches...without excluding other classical and scientific branches ...." Then, as now, not everyone in academe was pleased by support for science and technology. The Rev. Dr. McCosh, President of Princeton College, condemned the whole policy. Notwithstanding such opposition, the Land Grant Act helped greatly in the development of higher education in science and technology in the United States. By the end of the 19th century there were more than 100 educational institutions in the United States teaching engineering disciplines. In 1860 there were 30 graduates per year from civil engineering schools in the United States, but there were none in mechanical or mining engineering. The first university program in metallurgy in the United States was apparently started at Yale in 1855, and university mining studies apparently began with the founding of the Columbia School of Mines in 1867. By 1890 the annual production in the United States had reached 310 civil engineers, 370 mechanical engineers, and 35 mining engineers. The first Department of Ceramics was started at The Ohio State University in 1894. Readey has summarized the subsequent founding of ceramic engineering programs at other colleges in the United States. By 1906 there were five programs and by 1949 there were eighteen programs. Also important to the field of ceramics was the founding of The Bureau of Standards 1901. This agency of the U.S. government subsequently became the National Bureau of Standards and later the National Institutes of Standards and Technology as its responsibilities were broadened.

Paralleling the founding of specialized technical university departments was the development of specialized national technical societies. The first national technical society in the United States was apparently the American Philosophical Society founded in 1743, which had as its field the systematic study of practically any category of phenomena. It had distinguished founders indeed. Its first president was Benjamin Franklin, and its third president was Thomas Jefferson. There was great activity in the late 18th and early 19th centuries in founding local and regional technical societies. The founding of national technical societies in great numbers came later. Ralph S. Bates in his study of scientific societies in the United States found a total of three national societies (some called academies) in 1785 and a total of 36 in 1865. He considers that an era of founding specialized technical societies began in about 1866 although the roots of some important national societies are earlier. Relating to ceramics we note the founding of the American Institute of Mining and Metallurgical Engineers in 1871, the American Chemical Society in 1876, the American Institute of Electrical Engineers in 1884, the American Geological Society in 1888, and the American Physical Society in 1899. The founding of The American Ceramic Society in 1899 was thus part of an extensive process of founding specialized societies that began in the second half of the 19th century and continues today. The founders of The American Ceramic Society recognized the importance of science, technology, and art. The central purpose was the promotion of knowledge about ceramics for the public good. An affiliate focused on the specific role of ceramic engineers, the National Institute of Ceramic Engineers, was added in 1938. Later societies important to ceramics include The Mineralogical Society of America founded in 1919, the American Society for Metals founded in 1920, and the Materials Research Society founded in 1973. The founding dates of many societies important to ceramic professionals are listed in Table 1.

Table I. Founding Dates of Ceramic Societies and Related Organizations

Date	Society	Founding location
1871	American Institute of Mining, Metallurgical, and Petroleum Engineers (AIME)	Wilkes-Barre, PA
1876	American Chemical Society (ACS)	
1884	American Institute of Electrical and Electronics Engineers (IEEE)	
1888	Geological Society of America	Ithaca, NY
1891	Ceramic Society of Japan (originally YooKoKai)	Tokyo, Japan
1899	American Ceramic Society (ACerS)	Columbus, OH
1899	American Physical Society (APS)	New York, NY
1913	ASM, International (Originally the Steel Treaters' Club)	Detroit, MI
1919	Mineralogical Society of America	
1919	Deutschen Keramischen Gesellschaft	Berlin, Germany
1929	Belgian Ceramic Society, retitled in 1987 from the "Society to Encourage the Study of Glass, Cement, and Ceramics".	Brussels, Belgium
1948	Nederlandse Keramische Vereniging	Utrecht, the Netherlands

1949	Scientific Society of the Silicate Industry (Hungarian Ceramic Society)	Budapest, Hungary
1960	Sociedad Española de Cerámica y Vidrio	Madrid, Spain
1962	Svenska Keramiska Sallskapet	Stockholm, Sweden
1967	National Council on Education for the Ceramic Arts (NCECA)	East Lansing, MI
1973	Materials Research Society	State College, PA
1983	Groupe Français de la Céramique	Paris, France
1987	European Ceramic Society	Cambridge, England
1987	Academy of Ceramics	Faenza, Italy
1989	Polskie Towarzystwo Ceramiczne	Cracow, Poland
1989	Slovak Silicate Society	Bratislava, Slovakia
1995	Danish Ceramic Society	Copenhagen, Denmark

Trends in Ceramic Technology since 1899

The founders of The American Ceramic Society wanted a forum in which to discuss science as it related to ceramics. They believed not only in the value of science for itself but also that applied science had great value in improving the processing and properties of ceramics. A notable feature of the growth of The American Ceramic Society has been the role played by many people from scientific and engineering disciplines in addition to ceramic engineering. The leading spirit among the founders, Edward Orton Jr., was a geologist. One might suppose that, with the development of many ceramic engineering schools and the growth of a strong ceramic engineering community, the role of other disciplines would become minimal, but personnel from other disciplines remain a strong component of the Society. Other fields have their own technical societies and publications, but people in these fields whose work leads them to ceramic materials find the Society a very useful place to present their work and to interact with others interested in ceramics. Little can be done with ceramic materials unless they can be made with controlled composition, structure, and microstructure. The American Ceramic Society remains an unexcelled place for forefront work on ceramic processing and characterization of composition and structure. The role of the Society as a meeting place for many disciplines as they apply to ceramics mirrors an important aspect of the development of ceramic technology. This technology has not advanced just from developments within the ceramic industry. It also has been profoundly affected by events outside this industry so that any account of the development of ceramic technology should consider external factors.

Developments in the 20th century that stimulated progress in ceramics include advances in science and technology in general, the rise of new industries, calamitous advances in military technology, and concerns for health, safety, and the environment. It is useful to think of ceramics in the 19th and 20th centuries in terms of three major effects. First, the rise of many new technologies outside ceramics called for materials with new and improved properties and stimulated development of new ceramics for some of these needs. Second, advances in science and in techniques for characterizing materials assisted in the development of totally new ceramics as well as leading to improvements in the properties of existing types of ceramics. Third, production of traditional ceramics was much improved by the application of advances in mechanical and electrical engineering to the existing processes of ceramic production. Innovation often involves a combination of these effects, but separate consideration remains helpful as an organizational mechanism. We begin with consideration of the advances in science and technology outside ceramics that had great effect on ceramics.

Events near the turn of the century set the stage for a still-continuing period of unprecedented growth in science and technology, with each assisting the other in a synergistic relationship. Maxwell, building upon Faraday's work, enunciated his theory of electricity in 1865, and Hertz in 1888 proved the reality of the electromagnetic waves that this theory predicted. G. J. Stoney in 1874 identified the discrete unit of electricity and later named it the electron. J. J. Thompson showed that cathode rays consisted of streams of particles and so established the reality of the electron in 1895, putting in place the building block upon which electronics is founded. Milestones in the subsequent development of radio communication include Marconi's transmission of radio messages across the Atlantic in 1901 and De Forest's development of the triode vacuum tube and its use in oscillators and amplifiers by 1910. In another stream of scientific development Max Planck inaugurated the quantum theory in 1900 with his discovery of the quantum of energy. His successors erected a theoretical framework that would provide understanding of electronic devices and quantitative modeling of them as well as suggesting new devices. The seeds of the age of electronics, an era that would build upon the still-growing age of electricity, were thus planted. The electronics and communications industries that grew in association with 20th century advances in science, especially physics, placed new demands on materials and greatly stimulated innovations in ceramics. The development of the transistor in 1947 and the growth of the computer industry from the 1940s to the present day have spawned innumerable spinoff and satellite industries with special requirements for ceramics.

The automobile, truck, and specialty vehicle industries are so vast that they present attractive opportunities for ceramics in special roles. When the combination of performance, complexity, high technology, reliability in the hands of the public, durability, and affordability are considered, automobiles are remarkable achievements indeed. The use of ceramics in automobiles is worth considering in detail not only for itself but also to illustrate the extent to which specialty ceramic items are essential as components of some devices and systems not commonly thought about in terms of ceramics. The following list of ceramic applications in automobiles was provided by Gary Crosbie of the Ford Motor Company.

Glass used in external structures

- Shaped and tempered glass
- Laminated safety glass
- Conductive defroster line paste combinations (silver–glass frit)
- Halftone opaqueing/masking frits for appearance

Illumination

- Lamp glass and glass–metal seals

Mirrors

- Dimming night mirror (tilting and electrochromic)
- Antireflective coatings

Noise control

- Soundproofing mats (fiberglass)

Reinforcement

- Fiber mats embedded in polymer composite body panels
- Rocker cover panels reinforcement
- Intake manifold reinforcement
- Metal-matrix composite reinforcement in piston heads
- Brake pad and clutch/band surface materials

Thick-film resistors (ruthenates) and alumina substrate for signal conditioning

- Integral alternator regulator
- Mass air flow sensor

Engine control circuits
Antilock brake control circuits
Yaw sensor
Throttle position sensor

Engine sensors

Magnets for crank angle position sensing
Knock sensor (accelerometer lead zirconate titanate (PZT))
Thermistors (e.g., air charge temperature to control input)
Manifold absolute pressure sensor (Al_2O_3, glass seal, silicon)
Differential pressure sensors (Al_2O_3, glass seal, silicon)
Exhaust gas sensor (ZrO_2, Al_2O_3, glass seals, and protective coating)

Electronic modules

Heat spreaders (BeO, AlN, Al_2O_3 for ignition and other high power modules)
Direct bond copper (AlN, Al_2O_3) module assemblies
Metal-matrix composite Al–SiC composite baseplate (EV power electronics)
Ferrites ($MnFe_2O_4$ electromagnetic compatibility)
Varistors (ZnO–Bi_2O_3 module surge protection)
Resonator (PZT clocking digital electronics)
Integrated circuit–resonator combination
Chip resistors and trim potentiometers
Disk capacitors
Multilayer capacitors
Fuse supports (Al_2O_3—European)

Other electronic devices or features

Global positioning
CD-ROM reading optics
Cassette tape read heads
Radio tuning inductor cores
Backup obstruction - ultrasonic detection (PZT)

Spark plug

Insulating body (88–92% Al_2O_3)
Glass–metal graded seals in resistor-type spark plugs
Pressed powder seals

Thermo-switch insulators (micaceous supports for bimetal plates)

Silicon micromachined injector orifice plate

Pump seals

Water–ethylene glycol coolant pump (Al_2O_3, some SiC)
Seat lumbar support inflators (graphite)
High-pressure injector pumps
Apex seals (Wankel rotary engine—Si_3N_4)

Other wear parts

Injector wear parts
Roller followers (Si_3N_4—large diesel United States)
Tappet shims (WC–Co—Europe)

Permanent magnets in small motors

Filters

Engine air filter (component of filtration media)
Cabin air filter
Filter (air bag detonation, hold back sodium)
Regenerable diesel particulate filter

Exhaust system

Port insulation (aluminum titanate)
Cermet exhaust gas recirculation valve bushings (bronze–graphite)
Turbocharger rotors (Si_3N_4—Japan)
Fibrous exhaust gasketing (aluminosilicate)
Fibrous thermal insulation (aluminosilicate)
Thermal insulation (around close-packed exhaust)

Catalyst tumescent wrap (so it does not leak)

The computer and associated industries also have been very important to ceramics. A significant point here is that, in addition to direct use of ceramics in computing and communications systems, the use of ceramic sensors has been stimulated. The availability of substantial computing power on a relatively inexpensive chip has made it possible to build computer control into many processes, such as automobile engine operation. Such control depends on sensors and actuators. Ceramics have found many specialized uses of this type.

The hundred years under consideration were divided almost in half by World War II. Although World War I had important impact on technology, the overall impact of World War II and the following economic boom, first in the United States and then in Japan, Germany, and elsewhere, had much greater impact. Contrary to the experience after World War I, the United States did not return to "business as usual" but settled into the long-term cold-war struggle. Massive government support for research and for education as well as massive funding for military advances stimulated an unprecedented level of activity in scientific and technological development. The founding during World War II of the national laboratories of what is now the Department of Energy provided large centers of research and development that included research on ceramics. The establishment of the Office of Naval Research and the National Science Foundation made funding for research available on a competitive basis throughout the U.S. university system. The establishment of the federally supported Materials Research Laboratories at selected universities stimulated scientific and technological advances in materials and their applications. The accompanying boom in the civilian economy provided markets for advanced products, some of which used ceramics.

In the past 100 years (and particularly in the past 50 or 60 years), a diverse field of high-technology ceramics (electronics, magnetics, optics, sensors, computers, etc.) based on synthetic raw materials has grown up in parallel to the clay-based ceramic industry (whitewares, refractories, structural clay products, and glass) that continues to be based largely on clay and other bulk raw materials (silica, etc.). A development of great importance to traditional ceramics and even more so to high-technology ceramics is the solution of ceramic phase equilibria and the increasingly systematic compilation and use of this information in the discovery and processing of improved ceramics. Complete melting is important in the preparation of glass and the growth of single crystals from the melt. Partial melting is used in liquid-phase sintering of ceramics as with most traditional ceramics. Melting must be avoided or at least limited in the use of refractories for steelmaking and other high-temperature applications. The visual expression of Gibb's phase rule in phase diagrams presents in useful form the melting behavior and much other information vital to advances in ceramic processing and properties. The lead in developing techniques for high-temperature phase diagram determination was taken by N. L. Bowen of the Geophysical Laboratory of the Carnegie Institution as early as 1914. He and his group developed the quench method that required special platinum sealing techniques, special furnaces, and good temperature controllers. The development by the

Braggs and others of X-ray diffraction as a tool for crystal structure determination made it the primary characterization tool in phase-equilibria studies although petrographic microscopy remains an important supplement. Bowen and others, including G. W. Morey and J. F. Schairer, used these techniques to conduct a program of phase diagram determination and founded a school that spread to many other places with particularly important groups at The Pennsylvania State University and at The National Bureau of Standards. The latter group became the home of the international compilation and dissemination of phase diagrams that is now done under the auspices of The American Ceramic Society.

The development of the chemistry, thermodynamics, and kinetics of point defects in crystalline materials in the past 50 years is another major advance with many implications. This science underlies the understanding of matter transport in sintering and creep. Point defect science also has a major role in understanding electronic properties of insulators and semiconductors and so of most ceramics.

The increasing concern and government involvement in safety, health, and the environment in the 1960s had major impacts on industry, including the ceramic industry. In the United States, the Environmental Protection Agency and the Occupational Safety and Health Administration have been particularly important as have their equivalents at the state level. Practices in the handling of raw materials and manufacturing of ceramics have been modified in many cases to improve safety and health. Waste disposal has been modified to protect the environment. Ceramics have found many roles in improving safety, health, and environmental protection. Ceramic wear materials protect many types of processing machinery in handling environmentally hazardous substances. Ceramic uses include flue linings for high-heat conditions, ceramic filters for removing contaminates, woven ceramic fibers for handling and filtering dust-laden waste streams, and high-temperature bags for dust removal. Nonporous ceramic linings are used for collection and movement of gases and particulate matter. Porcelain-enameled metal sheets are used for corrosive gas exposure, and enameled vessels are used for reaction and mixing vessels. Special cements line collecting ponds and hold together bricks in kilns and stacks. Some cements are designed for air and liquid exposure and offer excellent service life.

Metals and alloys are most commonly made by casting from the melt although sintering of metal powders also is used to some extent. Containing the metal melt requires ceramic refractories. More than 50% of refractories are used in the iron and steel industry. Changes in the demand for types of steel and advances in ironmaking and steelmaking technology have caused great changes in refractories in the past 50 years as summarized by Marvin. Clay refractories based on alumina–silica fireclays were traditionally used. In blast furnaces for ironmaking and open hearths for steelmaking the traditional refractory-lined walls have been largely replaced above the melt by water-cooled metal walls. Requirements of the automobile industry for higher-strength steels free of refractory inclusions have led to the use of higher-quality refractories in contact with the melt. Traditional fireclay linings have been replaced by high-alumina, silicon carbide, and carbon. Silica or high-alumina is now used in blast furnace stoves. Steelmaking is now done mostly in basic oxygen furnaces (BOF) rather than open hearths. Magnesia combined with graphite or carbon is used as the refractory lining. The resistance of this lining in combination with new techniques of hot maintenance of the lining by flame gunning have led to lives of as much as 24,000 heats. Maximum throughput of steel in the BOF has led to final adjustment of composition in the ladle, which again required replacement of the fireclay refractory with high alumina or magnesia. Improved slide gates to control pouring of liquid steel have been developed that are made of high alumina, magnesia, and alumina–graphite.

In addition to looking at the effects of new industries on the need for ceramics, it is useful to look at the way in which advances in manufacturing technologies affected the production of traditional ceramics. The ceramic and glass industry at the turn of the century was greatly and favorably impacted by the general development of mechanical devices and electric power.

In contrast to metals, crystalline ceramics are most commonly made by consolidating powders (sintering), but melt processing also is used for glasses, glass ceramics, single crystals, and a small amount of so-called fused cast refractories. We consider developments in powder-based processing of ceramics first.

The major stages in the powder-consolidation method of ceramic production are production of the powder, forming of the unfired article, and sintering (firing). Traditional starting powders for ceramic processing are made from naturally occurring mineral deposits. Removal of iron by magnetic separation is important in making ceramics free from discoloration. Many high-technology ceramic products require synthetic powders. For example, alumina sparkplugs depend on high-purity alumina prepared in bulk by processes developed mostly in the 1930s.

Early ceramic-forming techniques were simple (although the resulting shape was sometimes complex) and depended on skilled hand labor. Traditional brick and tile were used as fired. In the 20th century complicated ceramic shapes with close tolerances as well as methods of mass production have increasingly been required. Machining of a fired ceramic is sometimes necessary but is held to a minimum because the high hardness of most ceramics makes mechanical finishing expensive. Accordingly most ceramics are formed and fired to produce a shape as close to the final shape as possible. This places special requirements on the forming and sintering stages. The earliest ceramics were made by hand-shaping plastic clay. Both the potters' wheel and plaster molds were used in antiquity. Contemporary forming techniques include slip casting, extrusion, uniaxial pressing, isostatic pressing, tape casting, and injection molding. Injection molding of plastics was developed in the 1920s and in mass production in the 1930s. Isostatic pressing of ceramics for spark plugs was introduced in the 1930s and is the predominant manufacturing technique used worldwide for such items today.

A major ceramic innovation in the late 1940s and the 1950s with large and still-developing applications is that of sheet forming of ceramics by the tape-casting process. Cutting and stacking produces multilayer assemblies that can be metallized to make multilayer capacitors or interconnected in three dimensions to form compact and hermetically sealed wiring harnesses for electronic chips. The history of tape casting has been summarized by Richard E. Mistler in the October 1998 issue of the *Ceramic Bulletin.* Glen Howatt built the first tape-casting machine at Fort Monmouth in about 1945 and published it with R. G. Breckenridge and J. M. Brownlow in the open literature in 1947. He received the first patent on tape casting in 1952. The process was further developed by knife coating the slurry onto a plastic film to create the tape-casting process widely used today. John Lawrence Park of American Lava Corp. in the early 1950s extruded thin sheets of alumina and other ceramics with an organic binder and plasticizer and received a patent in 1961. The process has undergone many detailed improvements including the use of better binders and plasticizers and the use of very fine powders. Warren Gyurk worked with Howatt at Gulton Industries and then moved to RCA Laboratories where he worked with Bernard Schwartz and Harold W. Stetson. Gyurk invented the lamination process for making multilayers for which he received a patent in 1965. Stetson received a patent for co-sintering metallized and laminated tapes into a monolithic structure in 1965. Gyurk and Stetson moved to Western Electric Engineering Center in 1963 where they developed a tape-casting process using fish oil as a dispersant to make aluminum oxide thin-film substrates for which they received a patent in 1972. Bernard Schwartz moved to IBM where he and his research team pioneered the multilayer ceramic technology for advanced electronic interconnect packaging applications. These advances led to very smooth, as-fired surfaces with a surface roughness of 1 to 2 microinches. This facilitates the application of conducting films. Patterns of electrically conducting lines were initially applied by screening-on pastes and later by sputtering films. Thousands of holes for location or pass-through conductors can by punched with a press. Two important classes of products resulted: multilayer capacitors and multi-layer ceramic electronic packaging. Both are made in the tens of billions yearly. This technology is still being extended to produce a new generation of products. Pioneering efforts include incorporation of portions of radio frequency wireless devices within the laminated layers.

Sintering (i.e., final compaction of the formed shape) of traditional ceramics includes the formation of a liquid phase during firing at high temperature that subsequently quenches to a glassy phase at room temperature. A major achievement of the mid 20th century was the development of a science of sintering that recognized several possible forms of matter transport including liquid-phase transport and transport by diffusion through the solid. Sintering is today an empirical process whose optimization for particular ceramics has been assisted by the understanding provided by sintering science. Sintering in the absence of a liquid phase is required for many high-technology ceramics. Such ceramics typically have substantial remaining porosity. Sintering science led to an understanding of how pores may become located within grains and cannot thereafter be eliminated. A triumph of ceramic technology in the 1960s was the development of the Lucalox process for sintering pore-free alumina and its extension to several other ceramics. Some covalent ceramics, such as silicon carbide, will not densify in the solid state under normal conditions. Another triumph was the development in 1973 of a process involving iron and boron additives for making dense polycrystalline silicon carbide.

Several major advances in the mass production of both traditional and high-technology ceramics took place in the 20th century with great progress being made in the 1930s through the 1960s. Improved methods of controlled agglomeration of powders, such as spray drying, facilitated rapid and well-controlled handling of powders. New forming techniques, such as dry-bag isostatic pressing, were introduced for mass production of spark plugs in the 1930s and were introduced into the manufacture of many other ceramic items. New, low-cost yet high-purity alumina (such as low-soda alumina) made possible mass use of alumina-based ceramics in many electronic applications where low electrical conductivity is required. New, sinterable powders, such as alumina doped with titanium or other additives made sintering as low as 1400°C possible and made mass production of alumina-based ceramics less expensive and more practical. Advances in tunnel kiln technology included the development of the roller hearth kiln with ceramic rollers, improved burner technology for lower cost and more uniform heating, and high-efficiency, lightweight fibrous thermal insulation. Together these tunnel kiln advances revolutionized manufacture and use of tunnel kilns.

Major components of the glass industry deal with bottles, flat glass, optical glass, and fiberglass. Many attempts were made to develop automatic bottle-making machines. Patents for various designs were issued in England beginning in 1859. In 1877 the semiautomatic Ashley machine was a commercial success in Yorkshire. The first fully successful automatic glass bottle machine was designed by M. J. Owens in the United States in 1898 and was in successful commercial production in 1904. A 10-head Owens machine could produce 2500 bottles per hour. The individual section (I.S.) machine, which uses gobs of glass from a feeder, was introduced in 1925 and gradually superceded the Owens bottle machine. The success of the electrical lighting industry in putting the Edison-Swan incandescent lamp into mass use depended upon the ability to make glass light bulbs cheaply. These bulbs were originally made by hand by glass blowers. In the late 1920s the Corning ribbon machine was introduced. A continuous sheet of glass is carried between opposed blowheads and blow molds. Glass bulbs are produced at a rate of 1800 or more per minute. Fluorescent lighting requires glass tubes. In 1917 Edward Danner succeeded in making tubing continuously by flowing molten glass around a rotating, heated metal rod and drawing it from the end. Glass TV tubes of optical quality are made by a special centrifugal process.

Flat glass was made in the late 19th century by three processes. In one, molten glass was rolled flat in a frame, cooled, and then ground and polished. In the second process, a cylinder was blown, cut along a line parallel to the axis, opened, and flattened. In the crown method, a sphere was blown, opened, and spun. In the 20th century a sequence of advances was made in forming sheet glass directly from the melt. The Fourcault process was patented in 1901 and operated commercially in 1913. Glass sheet is drawn upward from a melt through a slit and cooled. In the continuous plate-glass process of 1926, glass from a tank furnace flows to rollers. The resulting glass plate is then ground flat and polished. The next great advance in flat glass was the ***float glass process***[1] developed by Pilkington in the late 1950s in which molten glass flows from the tank onto a sheet of molten tin and is then withdrawn on rollers. Grinding and polishing are eliminated.

Glass fibers considerably predate 1899. Continuous-filament fiberglass is many centuries older than the glass wool type, but the latter was first made commercially in large quantities. Glass wool bats were used at least as early as 1840. There are many variants on the basic process of making glass wool using steam jets. Bats of glass wool came into large-scale production in the 1930s and are extensively used as thermal and acoustic insulation in the construction industry. Bats of glass fibers with compositions specially formulated for use at high temperatures can be used continuously at least as high as 2600°F. These insulating bats have revolutionized furnace design in the past 50 years. Fiberglass air filters for the home furnace are ubiquitous. Fiberglass also is used widely in industry for liquid filtration.

Hand-drawn fibers were known in antiquity. Drawing of single glass fibers was done at least as early as the 18th century. At the Colombian Exposition in Chicago in 1893 a model wore a dress made of glass fibers. These fibers had been drawn through a hole in heated platinum, wound on a drum, and woven into cloth. However, mass commercial use of glass fibers developed greatly in the 20th century, because the founding of the Owens-Corning Fiberglas Corporation in 1937 to mass-produce continuous glass fibers. The process involves drawing as many as 200 fibers simultaneously from a melt through a platinum bushing. The resulting fibers are widely used as reinforcing in polymer matrix composites and are found in many applications from boat hulls to aerospace vehicles.

Optical uses of glass fibers include both short-path and long-path communication. Bundles of continuous fibers are used over short distances for light and image transmission including such uses in medical devices for diagnosis and surgery. High-silica optical fibers with low loss (typically 4 dB/km or less) are now widely used for long distance communication. A recent development is the fiber optic amplifier.

The increasing use of composites in the 20th century is another trend important to ceramic development. In some cases two types of ceramics are used to make a composite (a so-called ceramic–ceramic composite), but more often a ceramic is combined with another class of material to achieve properties not available in a single-component material. Fiberglass-reinforced plastic is an example of a composite with extensive use in automobiles, boats, sporting equipment, etc. Other examples of important materials involving composites include magnetic tape, retroreflective optics, dental restorative materials, certain bone substitutes, multilayer ceramic substrates for electronic chips, composite piezoelectric transducers and actuators, and many advanced aircraft materials.

The evolution of ceramic abrasives and cutting tools constitute a major series of developments in ceramic technology that led to several successful ceramic industries. The Acheson process for silicon carbide in the 1890s established silicon carbide as an abrasive powder that is still widely used in grinding. The subsequently developed ceramic-bonded silicon carbide grinding wheel also is used widely. Synthetic diamond abrasive is used in diamond cutting saws and in polishing wheels. Another line of development is that of cutting tools for machining of metals and other materials. The cobalt-bonded tungsten carbide cutting tool is the standard of the industry. Both silicon nitride and whisker-reinforced alumina cutting tools have substantial specialty uses.

Advances in Ceramic Technology since 1899 Grouped by Applications

The principal content of this book is the collection of descriptions of ceramic innovations presented in the next section. We think it useful also to present here brief descriptions of selected areas of application and to highlight ceramic innovations that were developed especially for these areas. It is of course true that many ceramic innovations have found use in more than one area. The use assignments given here are sometimes not unique, and we have mentioned the same item in more than one category in a few cases. Even so, the groupings used here are useful in providing some degree of organization to the very diverse set of applications of ceramic innovations of the past century. The reader is invited to turn from an item mentioned briefly here in bold Italics to the detailed description in the next section.

Archaeology

Pottery was an early development of civilization. Archaeologists have found pottery very useful in dating strata being studied. Extensive tables of styles and dates have been developed and cross correlated with other means of dating, such as historic records when these are available. These techniques are increasingly being supplemented by dating based on technical factors, such as radiocarbon dating of organic artifacts. A relatively new technique useable with a tiny bit of a ceramic is ***thermoluminescence dating of ceramics.***

Automobiles and Related Vehicles

By weight the largest use of ceramics in vehicles is in the form of glass for windows. A recent ceramic innovation in this area is the ***automotive solar control electrically heated windshield.*** An early and essential development was the sparkplug. An innovation of the last century is the use of high-alumina ceramic sparkplugs that depended on the development of ***low-soda calcined alumina.*** Sensors have become very important in automobiles in recent decades for engine control and emission control as well as for various other roles. The ***zirconia oxygen sensor*** is today found on all automobile engines in the United States and is an essential part of efficient and low-pollution engine operation. Another essential part of pollution abatement through automobile emission cleanup is the catalytic converter using ***honeycomb ceramics.*** Small electric motors with ***hard ferrite*** magnets are found in applications such as power seats and door locks. As many as 20 such motors are used in some automobile models. In some diesel engines silicon nitride pre-combustion chambers and exhaust valves are used. Metal-matrix composites using ceramic fibers are coming into use for pistons and crankshafts. A ceramic wear-resisting seal is used in automobile water pumps.

A vast and complex system of roads, signs, and traffic control devices is required for use of automobiles and trucks. Good signs for all-weather viewing are provided by a polymer–glass composite using ***glass microspheres*** as optical elements as composite retroreflectors.

Ceramic Processing Technology for General Applications of Ceramics

As discussed previously, ceramic processing in the past century has been a major area of ceramic innovation. This technology advanced both through new techniques that bypass the use of powder consolidation and through improvements in traditional powder-based processing. Innovations in the latter include improved raw material that is discussed below as a separate category.

Mineral deposits have traditionally been used as raw materials with little or no beneficiation of natural raw materials. Magnetic separation has come into use to remove iron-containing particles that can cause discoloration. Synthetic powders (see, for example, ***low-soda calcined alumina, tabular alumina aggregates, calcined and reactive aluminas, tabular alumina aggregates, silicon nitride powder improvement, nanophase ceramics, nanopowder process for ultra-fine and ultra-fast pulverizing***) are now in wide use.

Powder-based ceramic processing was advanced by the development of free-flowing powders that facilitated automatic loading of molds. A major advance here was the development of controlled agglomeration to form a free-flowing powder by the ***spray-drying process***. Innovations in the area of forming the unfired ceramic include enzyme ***catalysis in ceramic forming, isostatic pressing,*** and ***ceramic injection molding***. A major advance in forming is sheet-formed ceramics and its development into ***laminated multilayer ceramic technology***.

Special chemical routes, such as sol–gel processing, sometimes combine aspects of powder preparation and forming into one process. Innovation descriptions include the early history of ***sol–gel ceramics, sol–gel processing of ceramics,*** and ***sol–gel ceramic products***.

Techniques for using applied pressure to facilitate sintering include hot pressing and ***hot isostatic pressing***.

Automatic machinery for ceramic firing advanced notably in the last half century. The roller hearth kiln led to very rapid firing of ceramic tile (as fast as 20 minutes) and of many other ceramic items. See ***kiln design innovations in the 20th century*** and ***conveyor technology for tile firing in tunnel kilns***.

Firing glass-free ceramics generally requires very high temperatures. ***Sintering of alumina at temperatures of 1400°C and below*** made possible a range of products. Glass-free ceramics were traditionally impossible to produce without some porosity. The development of ***pore-free ceramics,*** such as alumina for sodium-vapor lamp envelopes, was a major innovation as was the later development of ***pore-free silicon carbide ceramics.***

Ceramic Coatings Technology

Ceramic coatings are used for a variety of purposes including wear resistance, erosion resistance, thermal protection, control of optical properties, and electrical insulation. Important innovations included plasma-sprayed ceramic coatings, ***flame-sprayed ceramic coatings,*** thermal control coatings, sol–gel ceramic coatings, and oxide infrared reflective coatings for lamps.

Chemical and Structural Studies

The properties of ceramics depend not only on their chemical composition but also on their crystal structure and their microstructure. We note two developments in this area in the last century that have been of very great importance to all types of ceramics. These are X-rays in materials development and ***phase diagrams for ceramists.*** These differ from most of the other innovations in ceramics in not being products in themselves. They are certainly developments and are so important to ceramic products that we list them as ceramic innovations despite their special character.

Consumer Products

Ceramics are used in a wide variety of consumer products. Preeminent are the many types of glass windows made by the ***float glass process*** including the ***enhanced float glass process.*** The development of ***glass-ceramics*** and ***photochromic and photosensitive glasses*** led to many specialized applications. Major consumer products include ***radiant glass-ceramic cooktops and borosilicate laboratory and consumer glassware.*** Dishes made of ***laminated glass*** are in widespread use. Heating elements are also familiar to anyone who uses an electric stove although few may know that a ceramic insulating layer (see ***tubular sheathed heaters***) is essential to their operation. Another widespread use of ceramics that is probably unknown to most users is ***ceramics for paper.*** A special use of ceramics is in vibration control. ***Piezoceramic vibration control*** is now being applied to skis.

Defense (See also Electrical and Electronic Uses of Ceramics)

Defense needs and opportunities have driven many advances in ceramics. In the constant struggle between improved penetrators and improved armor, ***ceramic armor*** plays an important role as an adjunct to metal armor. Piezoelectric ceramics have many applications, but their development has been pushed especially by the need for improved ultrasonic submarine detection using piezoelectric transducers operating in the kilohertz range. Following the development of ***barium titanate*** in the 1940s by Wainer and Salomon, lead zirconate titanate (PLZT) was discovered at the National Bureau of Standards by B. Jaffe, R. S. Roth, and S. Marzullo in 1954 and in modified form remains the basis of ***piezoelectric ceramics from igniters to computers and telecommunication devices.*** Smart, heat-seeking missiles require infrared detectors, such as ***uncooled infrared cameras. AlON transparent ceramic (aluminum oxynitride)*** finds application as optical windows and domes, sight tubes, etc., where transparency and high durability are important. Eye protection for pilots and others utilizes ***electooptic ceramics.***

Electrical and Electronic Uses of Ceramics

The insulating, dielectric, piezoelectric, magnetic, and other properties of various ceramics have led to their widespread use in electrical and electronic devices. Some uses have already been described under defense applications; certain others are mentioned here. Electrical insulators (glass or porcelain) were used on the early telegraph lines. As semiconductor chip technology developed beginning in the 1940s, a special ceramic mount, used by the millions per day, was developed (see ***sintering of alumina at temperatures of 1400°C and below***). Another modern development is multilayer integrated ceramic packaging. This technology rests on ***sheet-formed ceramics, laminated multilayer ceramic technology,*** and metallization of ceramics including ***gas–metal eutectic direct bonding for advanced metallization and metal joining.*** Today virtually any ceramic can be laid down as a thin film, and thin-film conductors can be applied and cofired with the ceramic. An intricate, multilevel, three-dimensional composite of insulating ceramic and conducting metal can thus be produced to serve as a mechanical mount, a wiring harness, and a hermetically sealed container for electronic chips in many applications.

Functional elements in electronic devices include capacitors with ***barium titanate*** dielectrics. High-frequency radio communication requires ***microwave dielectric ceramics*** that are used as components of highly selective filters for cellular telephones and satellite communication systems. ***Soft ferrites*** are required for small but sensitive radio antennas and in ***magnetic recording media*** as well as in ***crystal-oriented hot-pressed manganese zinc ferrites*** for read–write recording heads. ***Hard ferrites*** are used in speakers and in many small electric motors including extensive use in automobiles as noted earlier. ***Zinc oxide varistors*** have become essential to protect electrical and electronic equipment from voltage surges from small overvoltages in electronic chips to huge voltage surges in lightening strikes on power lines.

The discovery of ***ceramic superconductors*** was a very exciting development of the 1980s that is still in its infancy.

Fluid Filters

Ceramics play a central role in important types of filters. Fiberglass filters for dust filtration are found in every home ventilating and air-conditioning system. Hot metal filters consisting of ceramics with controlled porosity are coming into increased use for the removal of particles from metal melts. ***Inorganic ceramic membranes*** have been developed with very fine openings. A low-cost, disposable ***ceramic water filter*** has been developed for field use by individuals.

Glass Production Technology for General Applications

Many glass items are made in enormous numbers at very modest cost by automatic mass production methods. Innovations in glass-manufacturing technology include the ***ribbon machine for glass bulbs,*** the ***float glass process,*** and the ***enhanced float glass process*** for making flat glass. Other important innovations are the ***Owens suction bottle machine, Danner process for making glass tubing,*** and ***continuous melting of optical glass.*** Glass-fiber-making techniques include those for ***continuous glass fibers*** and two techniques for fiber mats: ***steam-blown glass wool*** and rotary fiberizing.

Lasers

Lasers are used for many applications including communication, machining, and medicine. One of the first lasers invented (and one still in use) is the ruby laser: see ***glass and ceramic lasers.*** Large ceramic single crystals also are used as lasers including ruby and nickel-doped yttrium aluminum garnet. Titanium-doped sapphire is used as the basis for one type of tunable laser. A glass laser that is very important in producing short, very-high-energy pulses is the neodymium-glass laser: see ***glass lasers.***

Machinery in General

Ceramics are useful for low-wear, high-hardness, and high-stiffness applications. A growing application is in hybrid bearings (see ***first man-rated spacecraft application for silicon nitride ceramic bearings***). The combined strength, relatively low-density, and high-temperature capacity have led to ***silicon nitride turbocharger rotors***. ***Glass-to-metal seals*** are in widespread use to provide insulated access for electrical lines through metal walls. Very-high-strength parts can be made of ***transformation-toughened ceramics*** and ***fiber-toughened ceramics.***

Machining of Metals and Ceramics

Machining of metals and ceramics requires cutting and grinding tools of greater hardness and wear resistance than the material being worked.

Metal turning is done with hardened tool steel or with cobalt-bonded tungsten carbide for the most part (see ***cemented carbide***), but for specialized applications ceramic cutting tools ***including silicon-carbide-whisker-reinforced ceramics*** are used. Finishing of the harder metals is often done by abrasive grinding and polishing. ***Advances in abrasive materials technology in the past hundred years*** include the development and commercialization of the ***Acheson process for silicon carbide*** and the development of ***synthetic superhard materials: diamond and cubic boron nitride.***

Mechanical Property Enhancement

Ceramics can be made quite strong but are characteristically brittle. Mechanisms for imparting greater toughness to ceramics have been extensively studied and several have been quite successful. These are ***transformation toughening in zirconia-based ceramics,*** whisker-reinforced ceramics, and fiber-reinforced ceramics. The latter requires ***continuous ceramic fibers.*** A successful development using continuous fibers is ***tough silicon carbide-based thermostructural ceramic-matrix composites.*** An approach akin to whisker reinforcement is the use of processing techniques that produce internally grown long and thin reinforcing crystals. See ***in situ reinforced silicon nitride.*** Another class of ceramic composites with useful mechanical properties is ***melt-infiltrated silicon carbide ceramics and composites.***

Glass is often given improved strength by thermal tempering and ***chemical tempering of glass products.***

Portland cement is probably used in greater volume than any other construction material. Very impressive improvements have been made with ***high-performance reactive powder concrete*** that is just coming into use.

Medical and Dental Applications

Ceramics are increasingly used in medicine in diagnosis and as implants. ***Bioceramics*** are increasingly used as bone replacements, especially in Europe. A notable advance was the development of ***bioactive glasses, ceramics, and glass-ceramics.*** The bioactive glass allows direct bonding of living bone to glass and of living tissue to glass. Ceramic coatings are increasingly used on metal implants to facilitate bone growth: see ***hydroxiapatite coatings assist bone growth. Dental ceramics*** are very widely used; few older people are without a crown or bridge. Some 80% of these dental restorations are made of porcelain fused to metal (see ***dental restoration: porcelain-fused-to-metal***). Computer-assisted X-ray tomography has become an essential diagnostic tool in medicine. A recent advance in this technology is ***ceramic scintillators for medical X-ray detectors in CT-body scanners***. Another widely used medical diagnostic tool is ultrasonic imaging. Ceramics are used for the signal generator and pickup; improved performance is provided by ***piezoelectric ceramics in medical ultrasonic imaging.***

Nuclear Applications

Ceramics have played a major role in the development of nuclear energy as summarized by James R. Johnson. Best known is the use of ***uranium dioxide (UO_2) as a nuclear fuel*** in nearly every nuclear power reactor worldwide. Uranium dioxide pellets are incorporated in zirconium alloy tubes and serve to heat water to make steam. Uranium carbide microspheres encased in pyrolytic graphite have been developed as carbide nuclear fuel for fueling gas-cooled reactors, a few of which have been tested, and are candidates for the next generation of reactors. Many other ceramics have found applications in the nuclear field including refractories for processing uranium and plutonium and/or their compounds. Graphite and beryllium oxide have been used as moderator materials. Boron and rare-earth compounds have served as control materials. The planetary probes to Mars and beyond would not be possible without lightweight, long-lived, and efficient power sources. Thermoelectric generators fueled by the heat from radioactive decay of plutonium oxide have powered equipment for space exploration (see ***ceramic fuel for space exploration***). Nuclear power produces highly radioactive waste that must be stored for several half lives of the most radioactive isotopes. ***Nuclear waste glasses*** provide the means to contain these isotopes safely.

Optical Communications

Optical communication over long distances has been made possible by the development of ***low-loss optical fiber*** with very low optical attenuation per kilometer. Even with such low-loss fibers the signal must be amplified periodically. This amplification was done at first by detecting the optical signal, amplifying it as an electronic signal, and retransmitting it with a modulated light source. The recent development of ***erbium-doped fiber-optic amplifiers*** has made it possible to amplify the light itself. Manipulation of the optical signal is facilitated by devices made using ***ultraviolet-induced refractive index changes in glasses.***

Radio and Television (See also Electrical and Electronic Applications)

Soft ferrites provide cores for high-sensitivity reception antennas. Television tubes require excellent optical quality in the front face and high dimensional control. The development of mass-production methods for making optical-quality TV tubes has been essential to the vast worldwide use of TV. The recent development of ***large, flat glass TV tubes*** is a further important step.

Refractories for Metal, Glass, and Cement Processing

Refractories for steelmaking have advanced continuously, but at least three major innovations can be recognized. These are the ***basic oxygen process refractories*** (described also as ***resin-bonded magnesia–graphite refractories for application in basic oxygen furnaces***), ***refractory slide gates,*** and ***advanced refractory castables*** (described also as ***high-purity calcium aluminate cements***).

Needs of the glass industry led to the development of ***fusion-cast refractories*** also described as ***fused-cast high-zirconia refractories, fused-cast refractories for glass-melting superstructure application,*** and ***fused-cast refractories for molten-glass contact application.*** Other special refractories developed for the glass industry include ***dense zirconia refractories*** and ***chromic oxide refractories.*** Electric melting of glass was facilitated by the development of special electrodes described in ***electric melting of glass with molybdenum or tin oxide electrodes*** and ***refractory tin oxide electrodes.***

The making of metal castings using ceramic molds is an old process that has been much improved in finish, accuracy, and ability to make hollow metal parts by the development of the ***ceramic shell mold investment-casting process*** and ***ceramic cores for investment casting.***

Portland cement manufacturing was facilitated by the development of ***doloma zirconia refractories.*** Refractories for general use are often subject to thermal shock. The development of new ***low-expansion ceramics*** provides a route to minimize this problem. Furnace efficiency in general has been greatly improved by the development of fibrous insulation including ***refractory insulating fiber*** in crystalline form and in glassy form as described in ***steam-blown glass wool*** and ***rotary fiberizing.***

Sensors

A wide and growing range of ceramic sensors is used. Some have been mentioned under defense, electronics, and optical communication. Ceramic and glass sensors include ***positive temperature coefficient resistors, fiber-optic sensors,*** and ***zirconia oxygen sensors.*** A special class of materials not only senses but also responds as described in ***smart electroceramics.***

Single Crystals

The ability to grow large ***single-crystal oxide materials*** has led to many applications including watch bearings and jewelry. Diamond-like gems are made in very large volume at low cost as in ***mass production of refractory oxide crystals: cubic zirconia.***

Synthesis of Totally New Ceramic Materials

The important ceramic material silicon carbide was unknown until it was discovered during attempts to make diamonds. The ***Acheson process for silicon carbide*** made possible a large industry producing cutting, grinding, and polishing wheels and saws. A variety of synthetic materials are made for special functional requirements including ***barium titanate*** and ***lead zirconate titanate.***

Part II

Descriptions of Important Ceramic Innovations in the Past 100 Years

Organization of the Descriptions

The descriptions of important ceramic innovations that are briefly summarized in this book have been collected into five categories.

First, there are innovations in processing of predominantly crystalline ceramics. These innovations may have arisen to improve production of a specific product, but generally have broad application to many products. They have been grouped into innovations in (1) powders, (2) forming, (3) firing, and (4) special chemical processing routes.

Second, there are innovations in processing of glass. Most of these innovations are in glass-forming machinery and are associated with various shapes or forms of glass items. These have been grouped into innovations in (1) glass windows, bottles, and bulbs, (2) fiberglass, and (3) glass specialty items.

Third, there are innovations in ceramics used in the processing of materials including other ceramics. These have been grouped into innovations in (1) refractories, (2) metal processing, and (3) abrasives, cutting tools, and wear-resistant materials.

Fourth, there are innovations in ceramics having special properties that make them useful in electromagnetic, chemical, and mechanical functions. These have been grouped into innovations in (1) electrical insulators and dielectrics, (2) transducers and actuators, (3) magnetic ceramics, (4) optical ceramics and glasses, (5) sensors, and (6) structural ceramics.

Fifth, there are innovations in special applications not covered in the other categories. These have been grouped into innovations in (1) medical and dental ceramics, (2) nuclear and environmental ceramics, (3) other special ceramics and techniques, and (4) coatings.

More than one description is given for several innovations, such as sol–gel ceramics, the ribbon machine for making glass bulbs, glass-ceramics, basic oxygen furnace refractories, dental ceramics, and so forth. This has been done to show different points of view and illustrate different aspects. New technologies have often evolved in stages rather than in one place and in one creative act. Sometimes important incremental improvements have been over decades following the primary advance as in the case of the ribbon machine.

2

Basic Ceramic Processing

The dominant form of ceramic processing starts with powder, forms a green article, and consolidates this into the finished product by firing. One alternate process, special chemical routes, is discussed later in this section. A second alternate process, crystallization of glass, is discussed in the section on glass processing.

Important innovations in the powder-based process include innovations in powders, forming, and firing. Such powder innovations have been essential for many contemporary advanced ceramics. Specific improvements in the composition, particle shape, or particle size of alumina and silicon nitride powders are described. A general trend of the past two decades has been toward very fine powders. Such powders are difficult to handle. One solution has been the development of weakly agglomerated powders formed by spray drying. Spray-dried powders can be easily handled in the first stage of forming and then compacted by cold pressing before firing.

Forming is the stage at which defects are often introduced that persist through the firing stage and reduce the strength of the finished part. Major innovations in forming have minimized this problem and have facilitated mass production of complex shapes at reasonable cost. Isostatic pressing has made possible the routine production of spark plug bodies with close tolerances at practical cost. An enormously far-reaching innovation is that of sheet forming, also called doctor bladeing. The further innovation of metallizing and laminating these sheets has produced the multilayer ceramic technology that is essential to ceramic capacitors and electronic packaging.

Advances in kiln technology in the past half century have revolutionized firing for many ceramics. Ware now moves through a kiln on rollers rather than on a cart in many cases with great savings in time and fuel costs. These advances took place at about the same time in several countries. Two descriptions are given—one from a Russian viewpoint and one from an American viewpoint. Although these descriptions overlap somewhat, they also give different perspectives.

Subtle modifications in composition have been found that greatly improve the practicality of sintering and the quality of the final product. One breakthrough was the development of alumina compositions that could be sintered at 1400°C or less. Another series of breakthroughs was the development of special compositions and associated furnace atmospheres and heating schedules that produce pore-free ceramics.

Another development of great consequence is that of special chemical routes that in some cases combine the powder production and forming stages. Three descriptions of sol–gel processing are given that move from a general history to important work on organic precursors in the 1940s and 1950s and then to a range of commercial products.

Powders

Low-Soda Calcined Alumina

by George MacZura, Alcoa, retired

The importance of corundum (α-alumina) additions to vitrified ceramics was recognized during World War I. According to Walter H. Gitzen, editor of "Alumina as a Ceramic Material," more planes were then being lost due to spark plug insulator failures than were being shot-down in dog-fights. The free crystalline alumina in the vitrified spark plug insulators improved mechanical impact resistance and failures due to thermal shock resistance sufficiently to allow pilots to demonstrate their skills.

In the 1930s the automotive industry recognized that superior high-voltage insulation could be obtained by using relatively high (80%–95%) alumina contents to produce suitable ceramic insulators for use in high-compression-ratio engines. Sodium oxide (soda) at 0.5% Na_2O is the main impurity in manufactured alumina used for making calcined, fused, and sintered aluminas. Although high alkali content is satisfactory for use in compositions designed for mechanical applications, it is necessary to use material with the lowest soda content economically possible in electrical and electronic ceramics manufacture.

Low-soda (<0.15% Na_2O) alumina is particularly important for spark plug insulators, because they demand high, hot electrical resistance, and also in certain parts used for ultra-high-frequency electrical applications.

Fused alumina was the only source of low-soda alumina until the late 1930s, but the high cost of grinding, acid leaching to reduce the soda, washing, and drying limited its use in ceramics. With the need for acid leaching, sintered tabular alumina (0.02% Na_2O) was produced at lower cost and increased the use and applications of high-alumina ceramics in the 1940s. However, the cost for grinding the massive corundum grain to a fineness suitable for the manufacture of ceramics was still high.

In the mid-1940s, the soda in Bayer-processed alumina was reduced using small mineralizer additions. During calcination the submicrometer α-alumina crystals were grown to the range 2–10 μm. The loosely bonded calcined α-alumina crystals could be easily separated during grinding of the ceramic batch, rather than having to be preground similar to the fused and tabular aluminas. The Alcoa/Thompson patent dated May 3, 1949, improved the control of crystal size during calcination, making several crystalline-sized aluminas economically available to the ceramic industry. These low-soda aluminas greatly enhanced the use and development of electronic alumina ceramics after the late 1940s and provided a technology base for the future development of thermally reactive calcined aluminas.

Tabular Alumina Aggregates

by Robert A. Marra, Alcoa

Synthetic aggregates are used extensively in refractory applications because of their high-temperature stability. Tabular alumina or sintered alumina is a premium refractory aggregate used in brick, precast shapes, and monoliths. The name tabular alumina is derived from the large (50 to >400 mm) tabletlike α-alumina crystals that are characteristic of the product.

The first dense sintered tabular alumina was prepared by Thomas S. Curtis in his patented shaft furnace or converter in 1934. The continuous shaft converter allows a rapid sintering rate at high temperatures in the range of 1800°–1950°C. During the sintering process, pores are trapped inside the grains by a process known as discontinuous grain growth or secondary recrystallization.

Well-sintered tabular alumina is characterized by its large crystal size, high amount of closed porosity, and very little open porosity. This microstructure leads to low water absorption (desirable in monolithic refractory processing), extremely high thermal volume stability, excellent corrosion resistance, and high thermal shock resistance. On the other hand, over- and under-sintered tabular alumina contains a large amount of open porosity, which greatly reduces the grain strength and thermal shock resistance while increasing the water absorption.

The manufacturing process of tabular alumina mirrors that of a ceramic sintering process. The calcined alumina feedstock (high surface area with fine crystal size) is ground by continuous ball milling. The ground alumina is agglomerated by various processes such as ball forming, pan agglomeration, briquetting, or extrusion. The agglomerated material then is dried and transferred to the top of the converter where the agglomerates are sintered or converted to the tabular alumina structure. The sintered tabular material is then typically crushed, screened, and milled to the desired size fractions.

The primary application of tabular alumina is as an aggregate in refractory brick, precast shapes, and monolithics although its physical, chemical, thermal, and mechanical characteristics make it an ideal material for other applications. Other common applications include electrical insulators, kiln furniture, high-temperature catalyst supports, bed media for the filtration of molten metals, and molds for high-temperature investment casting.

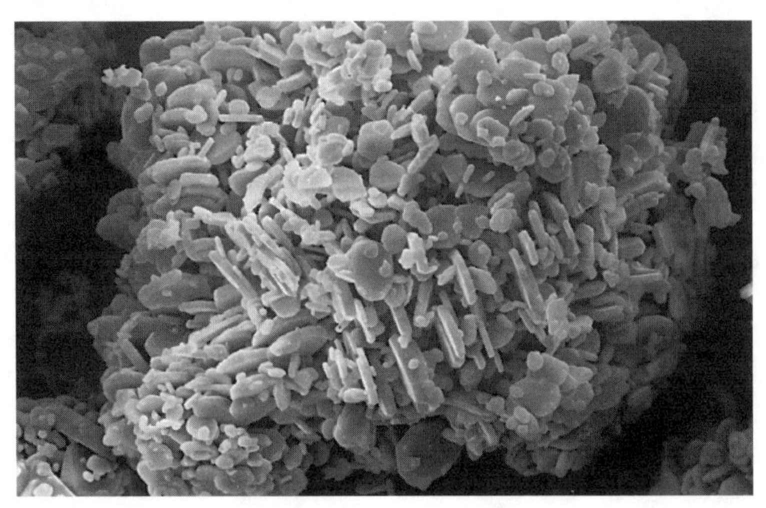

Figure 1. *SEM photomicrograph of Alcoa Industrial Chemicals unground calcium alumina. Courtesy of Alcoa.*

Figure 2. *SEM photomicrograph of Alcoa Industrial Chemicals reactive calcined alumina. Courtesy of Alcoa.*

Calcined and Reactive Aluminas

by Ray Racher, Alcoa

The timeframe of the development of calcined and reactive aluminas very much parallels the existence and growth of The American Ceramic Society. The development of chemical-grade aluminas is not recent, but still remains a very small part of the overall production of alumina chemicals. In the latter half of the 19th century, the development of the Bayer process for extracting alumina from bauxite opened up the potential of manufacturing pure aluminum oxide powders for a variety of applications, including advanced ceramics and refractories. Several refineries based on the Bayer process were constructed during the early part of this century. These Bayer plants (also known as bauxite refineries) use the Bayer process to extract aluminum hydrate from bauxite, leaving a red mud residue. Plants of this type are capable of producing very high volumes of aluminum hydroxide, which is then lightly calcined to make smelter-grade alumina to be turned into aluminum metal in an electrolytic process. Because of the availability of large volumes of relatively low-cost high-purity aluminum hydroxide and aluminum oxide, potential customers and operators of Bayer plants began exploring the potential applications of these alumina chemicals.

In the early 1920s ground calcined alumina was sold to abrasives producers and to polishing compound manufacturers. In the 1920s and 1930s, the addition of alumina to whitewares and electrical porcelains was developed, leading to the now considerable use of ground calcined alumina in spark plugs. Electronic substrate applications were first explored in the 1950s and now consume large volumes of very-high-purity aluminas. The market for alumina ceramics of 90%–99.9% purity would not exist today were it not for the availability of calcined alumina powders.

Only 7%–8% of the output of a Bayer refinery typically goes into the higher-added-value alumina chemicals (which include specialty hydrates as well as the calcined and reactive aluminas used in the ceramics industry). Of this, over half (60% is a recent estimate) is produced as activated aluminas and hydrated aluminas; therefore, <3% of the annual Bayer refinery output is sold as calcined alumina. The Bayer alumina hydrate used as the starting point for calcined alumina is essentially dry granules. These feedstocks are available in many forms with differing chemistry and surface areas. These products are fed usually to a rotary kiln, although other kiln types are used. The Bayer hydrate can be washed before calcination to remove some of the leachable soda. The goal is to convert the Bayer hydrate to aluminum oxide of essentially the α phase. This is achieved primarily by control of the calcination temperature. Mineralizers such as boron, fluorine, and chlorine can be used in the process to reduce the temperature of calcination, to produce alumina crystals of larger size, to remove additional soda by volatilization, or all of the above.

After calcination, the alumina then can be ground to produce a powder of controlled crystal size. This can be done in continuous mills or batch mills of various types. Ball mills are the most common pieces of equipment for size reduction, but vibrating mills and jet mills also are used for dry milling, and attritors for wet milling.

The variety of feedstock starting points, calcination routes, and grinding equipment means that there is a wide variety of calcined aluminas available to the marketplace. These are typically separated by chemistry (primarily soda content) and reactivity (measured by surface area, crystal size, ceramic properties, or a combination thereof). A typical range of calcined aluminas could be anywhere from the large, platelike crystals of polishing grade aluminas (up to 25 μm is not unusual) down to the submicrometer particle size of the aluminas intended to maximize ceramic properties such as fired density.

Figure 3. *Tabular alumina comes in many different forms: from tabular balls to coarse, crushed granular sizes, to fine ground powder. Courtesy of Alcoa.*

The availability of these different grades of calcined alumina is generally a result of the varying technical requirements of the different end-use markets. For example, spark plug manufacturers have typically desired a coarse crystalline alumina for good electrical properties as well as ease of manufacturing (specifically the grinding stage). Manufacturers of electronic substrates have been demanding ever-higher purity (particularly low soda) and good ceramic properties. Manufacturers of wear media demand good grindability, because this is a rate-determining step in their production process.

Applications in refractories have increased dramatically over the past few years with the development and introduction of low-cement and ultra-low-cement castables, where the fine particle size of the submicrometer reactive aluminas substitutes for calcium aluminate cement and fills voids in the castable that might otherwise contain water that would eventually leave a pore. Continued developments have led to multimodal reactive aluminas, where aluminas of two or more distinct particle sizes are ground together using particle-packing theory to optimize their rheological properties.

Additional sources of feedstocks for reactive aluminas are the various chemical processes used to make high-purity alumina such as that used in the production of sodium-vapor lamp envelopes, as well as biomedial applications. Further developments include the pilot production of nano-sized aluminas used in composite manufacture. These activities ensure a long and healthy future for alumina chemicals into their second century.

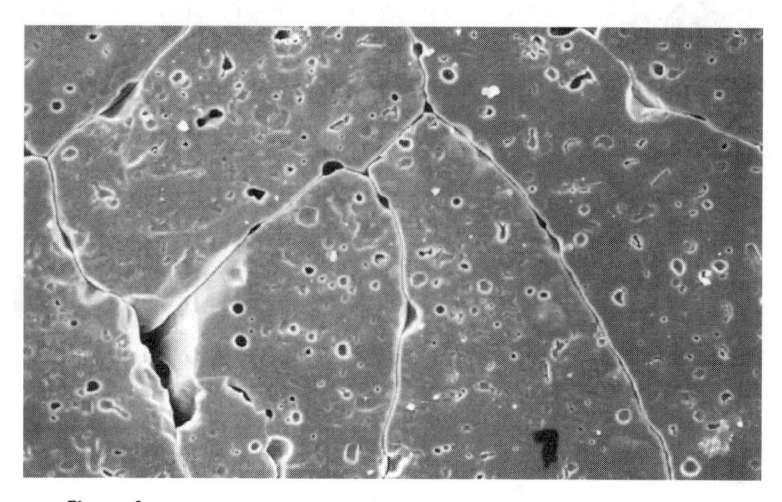

Figure 4. *SEM photomicrograph of a polished surface of Alcoa Industrial Chemicals tabular alumina. Courtesy of Alcoa.*

Silicon Nitride Powder Improvement

by Tetsuo Yamada, UBE Industries, Ltd.

Silicon nitride-based ceramics emerged in the 1960s as having outstanding potential for high-temperature structural applications. In the early stages of their development, silicon nitride ceramics were porous and had low strength. Powder designed for processing into high-performance ceramics has long been a key technology to fully meet functional requirements of engineering ceramics, because physical and chemical properties of silicon nitride ceramics are strongly influenced by the characteristics of the starting powder. Important properties of high-quality silicon nitride powder include high purity, small particle size, freedom from agglomeration, narrow particle-size distribution, and equiaxed particle shape. These characteristics of high-quality powder were not attainable by traditional techniques based on a solid-state reaction followed by crushing and grinding.

In 1982 UBE Industries developed a preparation process for high-purity silicon nitride powder by a chemical synthesis route. Silicon diimide is prepared by a proprietary liquid interface reaction method, wherein halosilane and ammonia are reacted at the interface of separated liquid ammonia and organic solvent. The resultant silicon diimide is decomposed to amorphous silicon nitride by calcination in a nitrogen or ammonia atmosphere after separating ammonium halides. The ultrafine amorphous silicon nitride powder is converted to α-silicon nitride by heating. The silicon nitride powder produced by this process has superior properties in terms of high purity, fine particle size, and high α content leading to the following advantages in sintered silicon nitride:

(1) Composition and phases of sintered material can be easily controlled because of the high purity.

(2) Highly dense material can be easily achieved because of the excellent sinterability of the powder.

(3) Uniform fine microstructures can be well developed because of the narrow particle size distribution.

(4) High-performance material with high strength and high toughness can be fabricated having a growth of needlelike grains caused by high α content.

Properties of Silicon Nitride Powders†

	Chemical composition (wt%)			
N	38.6	38.7	38.7	38.7
O	1.3	1.1	0.9	1.2
C	0.1	0.1	0.1	0.12
Cl	0.01	0.01	0.01	0.0
Fe	0.005	0.005	0.005	0.005
Al	0.002	0.002	0.002	0.002
Ca	0.001	0.001	0.001	0.001
α-Si_3N_4 (wt %)	97	100	100	98
Specific surface area (m^2/g)	11	5	3	7
Mean particle size, d_{50} (μm)	0.5	0.7	0.9	0.7

†These powders are widely used in making ceramics for highly stressed applications, such as cutting tools, bearing balls, glow plugs, and turbocharger rotors

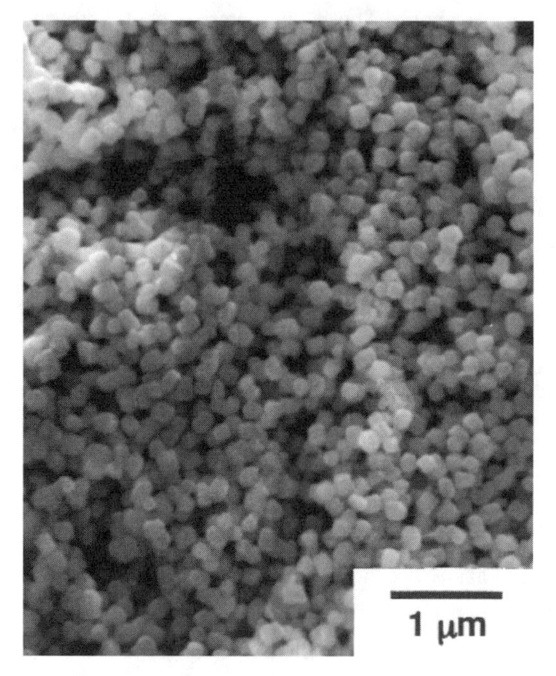

Figure 5. *SEM photograph of SN-E10.*

Nanophase Ceramics

by Bernard H. Kear and Ganesh Skandan, Rutgers University

Since the mid-1980s there has been growing interest in a new class of materials, called nanocrystalline, nanophase, or nanostructured materials, consisting of particles or grains of nanoscale dimensions (1–100 nm). The general term nanostructured refers to either powders, laminated structures, coatings, or bulk materials. Nanocrystalline materials, which have a very high surface or interface area, exhibit dramatic changes in properties, such as enhanced sinterability at low temperatures, improved ultraviolet light scattering, very high hardness and wear resistance, enhanced gas sensitivity, smaller particle size in colloidal suspensions, superior magnetic and dielectric strength, and enhanced optoelectronic properties. In order to form nanostructured coatings and bulk structures, nanopowders are used as the starting materials. Since the mid-1990s there has been an explosive growth in the number of applications for these types of materials as established technologies began to use nanostructured materials instead of coarser-grained materials, and new applications evolved to exploit the unique properties of these materials.

With increasing applications for nanocrystalline coatings and bulk materials, several powder production technologies have evolved to meet the demand, including wet chemical synthesis, spray pyrolysis, sol–gel processing, and vapor-phase condensation. Vapor condensation in a reduced-pressure environment has filled the niche for producing loosely agglomerated nanoparticles in large volume. Two scalable variations in this technology have emerged for the production of powders that consist of individual nanoparticles, as opposed to powders composed of nanograined large particles: (1) gas phase condensation (GPC), where a metal is evaporated at high rate in a controlled atmosphere, condensed into nanoparticles, and subsequently oxidized to produce oxide nanopowders; and (2) chemical vapor condensation (CVC), where vapors of chemical precursors are decomposed and condensed in a controlled atmosphere to produce a continuous stream of nanoparticles. In one variation of the CVC process, a flat-flame combustor provides the thermal energy for pyrolysis: combustion flame–chemical vapor condensation (CF-CVC). The flat-flame combustor ensures a uniform temperature, residence time, and precursor concentration across the entire face of the burner, thereby leading to a narrow particle-size distribution. The CF-CVC process is used to produce oxide ceramic nanopowders. In another variation of the process, a multihole hot-wall reactor is used to pyrolyze and condense the nanoparticles: (hot-wall–chemical vapor condensation (HW-CVC)). The HW-CVC process is used to produce non-oxide ceramic nanopowders. Powders produced by the vapor-phase method include SiO_2, TiO_2, Al_2O_3, SiC, Si_3N_4, and AlN.

Nanopowder Process for Ultra-Fine and Ultra-Fast Pulverizing

by Koicki Kugimiya, Matsushita Electric Company

"Rolling stones gather no moss," as the Chinese noticed thousands of years ago. This process is a very simple principle for pulverizing minerals to powders for ceramic processing. Ball milling has been a very traditional way to make raw powders for a long time and almost no big changes in the technology have been made until very lately.

Fine powders are necessary to get uniform mixtures and to lower sintering temperatures. Thus, many efforts have been conducted to achieve this goal.

A unique and economical process, the nanopowder process, that involves high-energy, high-speed, and ultrafine milling by a media agitating mill was introduced by Nishida, Kugimiya, and others in the late 1980s. In contrast to previous ball-milling processes, the new milling process utilizes smaller balls, higher agitating speeds with high energy, and extraordinarily high slurry concentration. The result is higher milling rates and lower contamination from balls and containers. A conventional ball-milling yields milling sizes of a few micrometers with days of milling time and unavoidable contamination. The new method yields submicrometer sizes within a few minutes and little contamination.

Taking advantage of this small particle size, low-temperature sintering is realized (150°C and lower), and even better properties are obtained because of the low evaporation of constituents and smaller grain sizes of sintered bodies.

The new method is now opening up new ceramic materials for various fields, including structure-controlled complex materials, layered electroceramics, and others.

Spray Drying

by Stanley Lukasiewicz, Texas Instruments, Inc.

High-speed powder presses require a feed material that will flow freely, uniformly, and reproducibly into the die cavity prior to compaction. Traditionally, ceramic press powders were prepared by wet mixing or milling bulky powder, followed by oven drying, crushing, possibly rewetting, and size classification. Unfortunately, this process was labor intensive, not easily reproducible, and did not produce a highly flowable, spherical granule.

The first discussion of using the technique of spray drying dates from the middle 1800s and the first spray-drying patent was issued in 1872. The first large-scale industrial use of the process was for milk and detergents in the 1920s. The chemical industry began utilizing it in the 1930s, and the food industry used spray drying to produce dehydrated foods in the 1940s. The ceramic industry began to utilize spray drying to prepare granulated press powders in the 1950s to reduce labor costs and to produce a higher-quality press powder.

Spray drying is an economical process that can produce large quantities of granulated powder directly from a ceramic slurry. A spray dryer operates by pumping a fluid feed material, typically the bulky ceramic powder mixed with water and possibly organic additives, to an atomizing device located in a drying chamber. There it is broken into a large number of small droplets, typically in the 20–500 μm range. The droplets, which tend to have a spherical shape because of surface tension effects, dry rapidly because of their high surface area-to-volume ratio. The solids that remain after the water has been removed from the droplets form the spray-dried granules. The properties of the spray-dried powder, which is composed of these dry granules, are controlled by the design of the spray dryer, the drying conditions within the chamber, and the composition of the slurry.

The empirical nature of the spray-drying process requires extensive testing with the particular ceramic slurry formulation prior to finalizing the design of the equipment. The size of the drying chamber, the type of atomizing device, the manner in which the atomized droplets mix with the drying air, and the method by which the dried granules are separated from the air and collected affect the efficiency and performance of the system. Atomization of the slurry is generally accomplished using a rotating disk or wheel, a pressure nozzle, or a pneumatic nozzle. The choice is based on the required drying capacity, the properties of the slurry, and the desired granule size. The airflow pattern within the drying chamber greatly influences the droplet drying time, residence time in the chamber, degree of moisture removal, and the formation of wall deposits.

In addition to the design of the spray dryer, the formulation of the ceramic slurry is a major factor controlling the final properties of the press powder. Proper selection of an organic binder (for nonclay-based compositions) produces granules that easily crush and deform during compaction while appropriate selection of the percent solids maximizes the bulk density of the spray-dried powder.

Since spray drying was introduced into the ceramic industry, it has become the most widely used technique for producing press powders. Properly done, it is an economical and continuous process that inherently produces a press powder that has uniform and repeatable properties.

Forming

Isostatic Pressing

by Jonathan W. Hinton, Lanxide Technology Corporation, and James S. Owens, Champion Spark Plug Company, retired

Ceramic component manufacturers have continually strived to increase the effectiveness of their shape-forming processes. The ability to produce a high and uniform green density is generally reflected in the quality of the finished product. Early ceramics were typically shaped by wet methods, such as slip casting, jiggering, or extrusion. Some damp forming and dry pressing were used, but shape capability was limited. The need for complex components with dimensional control, high strength, and low cost was realized with the advent of isostatic pressing.

Conceived, developed, and patented by B. A. Jeffrey at Champion Spark Plug Company in the 1930s, the isostatic-pressing process quickly became the preferred method for shaping spark plug insulators at Champion. The process was adopted by virtually all manufacturers of automobile and aircraft spark plugs when the Jeffrey patents expired. In fact, innovations similar to isostatic pressing have enabled spark plug makers to sell their products today for the same actual price as they did in the early 1900s. How many other products can you put on that list?

Isostatic pressing begins with fine but very flowable powders, generally prepared by spray drying or other methods of pelletization. A small amount of binder, which has been optimized for both green strength and green machinability, is added before spray drying. These powders flow quickly and uniformly into the pressing container, which is typically a rubber or other elastomeric bag of appropriate size and shape. If the desired article has internal details, as is the case for spark plug insulators, a rigid central mandrel is used, and the powder fills the annular space between the rigid mandrel and the flexible bag. The filled bag assemble is then subjected to a uniform external fluid pressure, to compress the powder. This "isostatic" pressure is generally applied with hydraulic oil in a sealed container. Typical pressures for isostatic pressing range from 4000 to 10,000 psi (28 to 69 MPa).

For low- to moderate-volume production, a "wet bag" process is used, with the bag assembly directly contacting the oil. This is typically accomplished by loading multiple filled bags into a large, thick-walled vessel, which is then sealed and pressurized. For high-volume production, which demands fast cycle times, a "dry bag" process is used. Dry-bag pressing utilizes a second, larger bag (often called the master bag) that is surrounded by the pressing fluid. This master bag transmits pressure to the smaller inner bag, which in turn compresses the powder. The inner bag remains dry, and can be quickly changed if it begins to wear, or to change to a new part number of different configuration. Dry-bag isostatic presses with multiple cavities can routinely produce 50–60 parts per minute.

The nature of isostatic pressing makes it particularly suited to low-cost, high-performance cylindrically or spherically shaped articles. Thus, its use has spread to a variety of shaped ceramic components, such as fluid flow nozzles, grinding media, and shaft sleeves.

To date, more than 50 billion components for spark ignition and similar applications worldwide have been produced by isostatic pressing. Despite advances with injection molding and other shape-forming processes, isostatic pressing remains unchallenged for low-cost, complex-shaped ceramic products as we approach the millennium.

Ceramic Injection Molding

by Gene Krug, Carpenter Certech

Ceramic injection molding is a marriage of two very different technologies: ceramic component manufacturing and plastic injection molding. The use of ceramic materials dates back to prehistory. Plastic injection molding dates back to the early 1900s, when plastic materials were first being developed. In the early 1930s, the marriage of these two technologies took place. This marriage was one of practical necessity. Namely, a manufacturing process capable of producing complex-shaped geometries, in a rapid, highly automated, and reliable manner was needed for the manufacture of ceramic components used in automotive applications.

One of the earlier uses of this process was in the manufacture of spark plug insulators. In his 1949 paper, Karl Schartzwalder describes the manufacturing process stages used for "Injection Molding of Ceramic Materials." These process stages included mixing, injection molding, binder removal, and sintering. These are the same basic processing stages used today. Schartzwalder explained that alumina powders (~60% by volume) can be made to behave in a plastic manner by the incorporation of thermoplastic binders. Up until this time ceramic engineers were only familiar with clay bodies exhibiting this characteristic plastic behavior. Alumina was traditionally known as a nonplastic body.

For the next several decades the use of ceramic injection molding was overtaken by other, now common, techniques, such as isopressing and green machining. The major pitfall of the ceramic injection molding technique was the discontinuous nature of the process. Namely, long debinding times (measured in days) rendered this a batch process rather than a continuous one. Therefore, the expectation that ceramic components could be molded in a highly automated fashion was not satisfied.

In the 1960s and 1970s the ceramic injection molding process gained some ground. During this time the Wiech method was developed. Using this method, a soluble binder phase was removed in an initial debinding step. This lead to the formation of pore channels such that the remainder of the binder (insoluble phase) could be removed more readily (i.e., faster) using traditional thermal treatments. This was a marked advancement. However, the technology remained based upon a batch-type process, whereby debinding remained a rate-limiting step and limited to rather thin cross sections.

In the 1990s several major plastic injection molding feedstock suppliers (AlliedSignal, Bayer, BASF, Hoechst, and Rohm and Haas) developed ceramic and metal injection molding feedstocks. The goal was to provide technological advances that had been lacking since ceramic injection molding's inception, namely, to provide feedstock materials premixed and ready for injection molding, and to provide ceramic injection molding process technology whereby complex-shaped components could be injection molded, debound, and sintered in a continuous fashion, with competitive dimensional and yield capabilities.

Today, ceramic injection molding is on the verge of being competitive with more traditional forming methods. The rule of thumb, "If you can press it, don't injection mold," will apply for some time. However, for complex-shaped geometries, ceramic injection molding is becoming an alternative route. The next decade should be very interesting to watch from a technological standpoint, one that may change centuries of ceramic forming tradition.

Enzyme Catalysis in Ceramic Forming

by Ludwig J. Gauckler, ETH Zurich; Thomas J. Graule, Katadyn Product AG; and Felix H. Baader, Tegimenta AG

Forming ceramic green compacts is performed by many different techniques today. The selection of the appropriate technique is governed by the geometric shape, the number of pieces to be manufactured, and the chemistry of the ceramic. All of today's processes introduce process- or starting-material-inherent imperfections in the green part, leading to inhomogeneous microstructures in the sintered parts. This results in a broad distribution of properties of the sintered body.

In this work, we describe a new ceramic green body fabrication method that allows the production of complex-shaped parts of high strength and reliability.

The method uses an aqueous, electrostatically stabilized ceramic suspension of low viscosity. It is cast into a mold and then coagulates, forming a stiff, wet green body. Coagulation is performed by changing the pH of the suspension and/or by creating salt directly inside the suspension using a controlled, time-delayed reaction. Enzyme-catalyzed reactions that are decomposing a substrate, or self-propagating decomposition reactions of a substrate, can be used. After the coagulation reaction in the suspension, the wet green body shows good mechanical properties and can be demolded, dried, and sintered.

This new forming process DCC (direct coagulation casting) is characterized by near-net-shape capability of complex-shaped parts with high quality and homogeneity in the green as well as in the sintered state. Only small amounts of organic additives, which amount to <0.5 wt% based on ceramic powder, are needed for the catalytic reactions. Because the forming process takes place without pressure and at ambient temperatures, inexpensive molds and tools can be used. Ceramic bodies with homogeneous, low-defect microstructures can be cast using almost any ceramic powder. Complex-shaped components with thin and thick cross sections can be cast because there is no gradient in density and/or temperature during consolidation of the green body. Alumina parts of average bend strength (680 MPa) and high reliability (Weibull modulus $m = 40$) have been demonstrated.

Tape Casting

by Eric R. Twiname and Richard E. Mistler, Richard E. Mistler, Inc.

The first published material about tape casting was by G. N. Howatt, R. G. Breckenridge, and J. M. Brownlow. It appeared in the *Bulletin of The American Ceramic Society* in 1947. The first patent, No. 2,582,993, was issued to G. N. Howatt in 1952. Glenn Howatt is typically named as father of this process, because the first patents and publications bear his name. There was a need to replace mica in capacitors, and tape casting was developed as a process to make thin, flat sheets from ceramic powders. This process, also known as knife coating or doctor blading, is a fluid forming method that meters a fluid suspension of ceramic powder and other additives into a thin coating which, when dry, is strong and flexible.

The formation of a basically two-dimensional layer of ceramic is useful in many areas. Tape casting is used somewhere in many, if not most, electronic ceramics processing. Single-layer substrates for thin-film electronics are tape cast. Thick-film applications, although roll compacted now, began through tape casting. The essentially two-dimensional tape also lends itself to stacking of one layer on top of another. This ability led to the development of multilayer ceramic packages (computer chips), multilayer ceramic capacitors (MLCCs), and printed circuit boards. Although many of these products now are made with plastic layers, their birth was in the tape-casting industry.

Tape casting allows the manufacture of smaller and lighter components. This ability lends itself to the current consumer electronics market. Tape casting is so widespread in this decade that it is often taken for granted. Some of the items used daily that contain components made with tape casting include telephones, pagers, computers, printers, keyboards (computer and musical), microphones, TVs, VCRs, stereo components, antennas, cars/motorcycles, airplanes, helicopters, cameras (disposable through video), coffee makers, watches, kitchen appliances (ovens through refrigerators through toasters), irons, missiles, spacecraft, sonar/radar, MRI and X-ray machines, golf carts, metal detectors, photocopiers, moon rovers, electric toothbrushes, lithium batteries, some Christmas ornaments, hair dryers, video games, overhead projectors, E-mail routers, and many rapid prototyping processes.

The consumer market continues to require smaller and lighter pagers and cell phones, and tape casting has become an invaluable process to meet the need. High-density circuitry for smaller, faster computers with more memory requires tape casting. Thin ceramic layers for high-wear areas use tape casting.

Tape casting has become a manufacturing method for countless thousands of electronic, aesthetic, and structural applications. Tape casting has been and will continue to be a powerful and successful ceramic innovation.

Sheet-Formed Ceramics

by Jim Thompson, 3M Company, retired

As electronic devices evolved from vacuum tubes to solid state, the printed circuit board was created. As electronic technology became more sophisticated, the board became more than an interconnecting device for components: it became an active part of the circuit with resistive paste being screened and fired on it. The board then needed to be ceramic in order to provide both the electrical insulating properties and thermal properties to withstand the manufacturing process.

Sheets of ceramic, typically 95%–96% Al_2O_3, were required, ranging in size from <1 in.^{2} to ≥2 in.^{2}, with a thickness of 0.010–0.050 in. Early attempts to make the substrates were by powder compaction (dry pressing). For high-volume manufacturing at a reasonable cost, this technique was restricted to the smaller areas, <2 in.^{2} and higher thicknesses, usually ≥0.030 in. because of the extreme fragility of the unfired substrate.

The invention that significantly changed the ability to manufacture large, thin substrates was the sheet-forming, or tape-casting process developed in the early 1950s and patented by G. N. Howatt in 1952 (No. 2,582,993) and by John Lawrence Park in 1960 (No. 2,966,719) and assigned to American Lava Corporation, which had been acquired by 3M Company in 1954. By 1958 this process started to yield a significant amount of salable product. Although the patent refers to the extrusion of a thin web with plasticized organics as the binder, this approach was limited to ~2 in. wide and 0.040 in. thick. By varying the binder and plasticizer, using volatile solvents in place of water and knife coating the slurry onto a plastic film, the tape-casting process was innovated and became the basis of a commercial application many orders of magnitude greater than the original patent. This eventually allowed the forming of continuous sheets up to 7 ft wide and as thin as 0.003 in. The web was slit into strips and fed into high-speed, metal-stamping presses for the shaping of small parts with precisely located holes, similar to phenolic insulators or circuit boards. The fired size could be accurately predicted to ±0.5% by controlling the particle-size distribution, binder content, and drying speed. This gave substrate hole location accuracy that would accommodate mechanized component placement.

The primary binder was polyvinylbutryal (4%–6% by weight), which, at that time and for many years later, was the organic material between the layers of glass in automobile safety glass. The plasticizer was Ucon oil (0.5%–1% by weight), and the solvent was toluol and methyl ethyl ketone in various proportions, depending on drying systems. In the early stages of the process, acetate film was used, but, because of the tensile strength demands of coated webs moving through long drying ovens, the film of choice became polyester, such as DuPont's Mylar™.

As the electronics industry moved from thick-film, screened-on resistive pastes to thin, sputtered films of tantalum, the surface finish of 25–40 μin. on 95%–96% alumina substrates was far too rough to be used. The next major innovation was substrates with an as-fired surface finish of 1–2 μin. This was accomplished with 99.5%-99.7% alumina body compositions made from 1 μm to submicrometer alumina powder, binder, and solvent systems that accommodated the considerable increase in surface area and mechanized, solvent atmosphere drying ovens that could eliminate drying cracks. It also was important to use a polyester film with an ultrasmooth surface finish that the tape could replicate. The net effect was a substrate with the required surface finish and accurate hole placement. Although some of this product was simple blank substrates 4–6 in.^{2} or more, much of it was used to produce a popular circuit board for Western Electric of 3.75 in. × 4.5 in. with up to 1000 accurately placed holes. Because an individual stamping die would cost $20,000 or more (1965 dollars), batch-to-batch shrinkage control had to be maintained within ±0.25%. This innovation was a key factor that allowed the telephone systems of the nation to move from the old rotary switching systems to electronic switching systems.

The primary commercial application through the years was alumina; however, the process also was used for titania capacitors, steatite insulators, and other ceramics.

Laminated Multilayer Ceramic Technology

by David L. Wilcox Sr., Motorola

The fabrication process entailing dispersing ceramic powders in a slurry containing a plastic binder and then casting this slurry, utilizing a "doctor blade" technique, to form green sheets has created a quiet revolution in the creation of high-performance multilayer ceramic devices and an industry producing multilayer-electronic-ceramic-enabled systems with revenues in excess of 20 billion dollars annually. Credit for this basic thrust belongs to G. N. Howatt (Patent No. 2,582,993 in 1952) and J. L. Park Jr. (Patent No. 2,966,719 issued in 1961).

This technology was further advanced by Warren J. Gyurk of RCA, who conceived the idea of laminating these sheets, reduced it to practice, and received U.S. Pat. No. 3,192,086 in 1965. A team at RCA, including Warren Gyurk, Harold Stetson, and Bernard Schwartz, used this technique to create the first multilayer ceramic capacitors, and they presented this work at an Electronics Division meeting of The American Ceramic Society in 1961. Patent No. 3,189,978 was issued in 1965 to H. W. Stetson of RCA on cocentering of metals and ceramics. Later, Bernard Schwartz and his research team at IBM pioneered the multilayer ceramic technology for advanced electronic interconnect packaging applications. Through careful matching of ceramic dielectric and metal conductor materials systems, advanced-high performance packages containing >100 m of imbedded metal conductor lines and >350,000 metallized vias allowing electrical connection and communication between the 30 or more layers are common today. Products using these structures are currently being utilized in IBM's mainframe and workstation computers and similar products offered by other computer manufactures.

Applications of the multilayer ceramic fabrication technology are moving beyond the original multilayer ceramic capacitor (MLCCs) and the multilayer ceramic electronic packaging (MCM-C) applications. Much as the thin-film multilayer technology has enabled the semiconductor industry to achieve the high level of component integration required for a variety of products, such as microprocessors, microcontrollers, and power amplifiers as well as MEMS devices. The multilayer ceramic technology is, in an analogous manner, enabling a variety of advanced products made possible by the higher levels of component integration, utilizing the unique electronic properties of advanced ceramics. For example, several companies are pioneering the use of the technology to integrate radio frequency (rf) wireless radio functions within the multilayer ceramic structures. The unique microwave properties of selected ceramic systems and the low losses associated with the integratable metal systems enable cost- and performance-effective integration of many of the rf portions of the wireless radios, leading to what might be called multilayer ceramic integrated circuits (MCICs).

The impact of the ability of this multilayer ceramic technology to create advanced, highly integrated ceramic devices that utilize the rich menu of ceramic properties is in its infancy. For example, the technology is being utilized to fabricate complex multilayer ceramic structures needed for current and advanced exhaust-gas emissions detection. It is an important technology for fabricating multilayer ceramic solid-oxide fuel-cell structures and is being explored for the fabrication of microchannel fluidic devices. Indeed, its impact on the advanced ceramic device industry may be viewed as analogous to the impact the thin-film technologies have had and are having on the semiconductor device industry!

Firing

Kiln Design Innovations in the 20th Century

by Robert J. Eicher, Swindell Dressler International Company

The recuperative muffle tunnel kiln, although inefficient by today's standards, was perfected in the early 1900s and was, in general, used for about 50 years until being replaced by direct-fire, low-thermal-mass tunnel kilns. Tunnel kilns, using the inherent advantages of a continuously operating heat exchanger, became common place in the first half of the century.

Swindell Dressler built their first metal roller hearth kiln in 1937 for decorating china. Higher temperatures were achieved in the 1960s with the use of ceramic rolls. Their roller hearth kilns were made to fire china, ferrites, pencil leads, wall and floor tile, stove tops, grinding wheels, substrates, and other products. Although roller hearth kilns are now made by many kiln suppliers, Swindell Dressler also has focused on low-mass kiln car technology, where firing cycles and production rates can equate with roller hearth designs without having the high replacement cost of rolls.

Figure 6. *Low-thermal-mass kiln-car designs by Swindell Dressler, with improved firing techniques, make wide-load, low-set firing of ceramics possible. Courtesy of Swindell Dressler.*

Major improvements in kiln technology appeared, however, in the second half of the 20th century. The first innovation occurred in the 1960s with the introduction of the high-velocity burner. The application of a high flame velocity recirculates the kiln atmosphere and leads to improved product temperature uniformity. This development allowed major reductions in firing cycles and meant a decrease in fuel consumption. This technique was quickly adopted by many brick, refractory, and whiteware manufacturers.

The next major innovation was the introduction of low-thermal-mass ceramic fiber. This material is now used in tunnel and shuttle kilns for all industries. It allowed kiln designers to reconfigure car loads and reduce load weights. Kiln designers could now build transportable, prefabricated kiln sections. This has helped to decrease field erection time, getting the kiln to the manufacturer quicker and cheaper. The use of ceramic fiber also helped increase product uniformity. Firing cycles could be reduced once more. Firing cycles of 30–60 h common to early 1900s tunnel kilns were now down in the range of 5–20 h.

The 1980s found the industry switching to computer-based control of all firing processes. The use of the computer continues to be perfected. The relative cost of computer control is probably the best bargain for a new kiln installation with reduced field wiring, more operational feedback, and quicker response to demand changes. Computer controls have allowed pulse-firing techniques to emerge into kiln design. Pulse firing puts the heat where and when it is needed. It made possible low-oxygen, low-temperature holds and ramps rather than high kiln temperatures being seen immediately when starting a burn.

Conveyer Technology for Tile Firing in Tunnel Kilns

by Anatoly E. Rokhvarger, Nucon Systems

Until the 1960s, tile and other thin-walled ceramic products were generally fired in tunnel kilns using a stacking technology that required a long process time. The method used was a batch-stack setting of clay brick, pottery, tile, sanitaryware, and china on the hearth or bottom of chamber kilns or on moving cars of tunnel kilns. Increased productivity was achieved only by increase in hearth capacities and kiln sizes or by increasing the number of kilns with a consequent significant increase in labor and capital costs. The traditional process requires several manual setting and charge–discharge operations. Also, there are unavoidable thermal stresses across stack setting that cause a high ceramic-scrap percentage. Moreover, high-energy input and a large amount of costly refractory brick are needed for large kilns.

It was realized that superfast firing of thin-walled ceramic materials could be achieved only by using a one-layer setting. In the middle 1960s, Efim L. Rokhvarger, A. B. Khizh, Ya. B. Kirshenbaum, and others developed improved industrial conveyor lines that included slot roller dryers and kilns for manufacturing wall, floor, and façade tile.

A typical processing time for tile of 20–60 h in a traditional tunnel kiln is reduced to 40–60 min in a one-layer roller hearth kiln. For the same production capacity, there also is a reduction in manual labor and in required floor space. The thermal efficiency of traditional kilns used in the tile, pottery, and whiteware industries is ~10% in contrast to 30%–45% achieved in roller hearth kilns.

The high productivity, cost efficiency, and other significant advantages of conveyor technology have caused a growth rate for tile production unprecedented in the traditional ceramic industries. By 1970 the former Soviet Union had 28 active conveyer lines that produced 5.4 m^2 of tile

Figure 7. *Conveyor line for wall tile. Courtesy William Kalinovsky, Kuchino Ceramic Plant, Moscow, Russia.*

per year. In Italy, the export of tile increased from an insignificant value in 1970 to $1.4 billion per year in 1987.

Conveyor technology is now a well-established technology for various thin-walled products, such as wall, floor, and façade tile, sanitaryware, pottery, and china. In Europe and Asia, most tile plants and a part of the whiteware plants have replaced their batch-stack car tunnel kilns with conveyor technology kilns. Significant replacement is also taking place in American ceramic plants.

Hot Isostatic Pressing

by Dale Niesz, Rutgers University

Hot isostatic pressing (HIPing) was invented at Battelle's Columbus Laboratories in 1955. HIPing is a process for applying isostatic pressure to materials at a high temperature. Typically this is achieved by placing a furnace inside a cold-wall pressure vessel. The process was first developed to hermetically seal UO_2 fuel elements in a metal (Zircalloy) cladding by diffusion bonding. The process was successfully applied to net-shape consolidation of beryllium powder in 1964. In the early 1970s, HIPing began to be used to consolidate powders to fabricate billets of high-speed tool steel and superalloys and for net-shape fabrication of structural titanium components directly from powders. In the mid 1960s HIPing began to be used to heal defects in aluminum, titanium, and superalloy castings. This process has gained wide acceptance for post-HIPing of aerospace metal castings, such as turbine blades.

Early work on densifying ceramic powders was done at Battelle in the early 1960s. The major advantage of HIPing of ceramics is the ability to apply high, isostatic pressure and high temperature simultaneously to densify a complex shape. This enables material to be densified that cannot be densified by sintering or uniaxial hot pressing and the densification of shapes that cannot be hot pressed. For many materials, it also enables the fabrication of higher-density, finer-grain-sized materials than can be densified by sintering or hot pressing.

There are two major methods for HIPing of ceramic materials. One method requires a gas-tight cladding or "can" and the second method does not. The cladding method is required if the material contains open porosity, because the cladding is needed to form a gas-tight barrier that allows the pressurized gas in the autoclave to exert a compaction force on the material. Either metal or glass claddings are used. Because a loose powder fill of most ceramic powders is usually too low in density to perform net-shape densification by HIPing, components are usually formed by some standard green-forming process before encapsulation. Silicon nitride bearings are densified by this method. The major disadvantage of the process is the cost associated applying the cladding, especially for complex shapes. This disadvantage is greatly reduced by a cladding process developed and patented by ABB Cerama in Sweden. In this process, the cladding is applied as a glass powder coating that is sintered to form a dense cladding during the HIPing cycle before pressure is applied. This process made it feasible to use the cladding method to HIPing complex shapes, such as the silicon nitride radial turbine rotor.

The second method of HIPing does not require gas-tight cladding, because the material is first sintered to the closed-pore stage before HIPing. This method often is referred to as post-HIPing. Because this method eliminates the cost associated with cladding, it is the preferred method for materials that can be sintered to the closed pore stage. The sintering step can be done prior to the HIPing cycle or incorporated into the HIPing cycle by delaying the application of pressure until the point in the cycle where the material reaches the closed-pore stage. This cladless HIP method is used for cemented carbides and soft ferrites, where performance requirements dictate premium properties.

Gas-pressure sintering can be considered a low-pressure version of post-HIPing. This process is often used for densifying silicon nitride, by sintering to the closed pore stage at a nitrogen pressure sufficient to prevent decomposition of silicon nitride and then applying a pressure of ~20 atm (2 MPa) to achieve further densification.

Sintering of Alumina at Temperatures of 1400°C and Below

by Edmond P. Hyatt, EPH Engineering Associates, Inc.

Before the advent of reactive aluminas (in the 1960s), which made practical use of high alumina as a stand-alone ceramic body, the only alternative was to fire alumina, with or without additives, to 1700°C. Independent of the efforts to provide grain-growth inhibitors and other additives, a team at the University of Utah, including this author, under the direction of Drs. Carl J. Christensen (formerly of Bell Telephone Laboratories) and Samuel S. Kistler (formerly of Norton Company) investigated the problem. After rather exhaustive experimentation, this team came up with the use of small amounts of titania and one or more of the transition elements to achieve densities of 3.8 g/cm^3 or higher when fired at temperatures as low as 1400°C.

A preferred composition was ~2% titania and 2% manganese oxide added to 96% high-grade alumina. This produced a dark, almost black, body but had very little deleterious effect on the strength and other desirable properties of alumina. The lowered firing temperature produced another benefit, that of reduced grain growth compared to the higher-fired aluminas. Although copper, nickel, iron, cobalt, and chrome oxides were attempted alone, the use of titania with these oxides was essential to obtain desired results. Combinations of these oxides produce opaque bodies of varying dark colors ranging from black to browns and even a purplish-brown. It was postulated that the added oxides really became sintering aids before any chemical interaction with the alumina. The observation that bodies, so made, "bleed" into setter plates lends credence to this idea.

This superior method of sintering alumina, even when used with the now-available reactive aluminas, has greatly expanded possible applications of sintered alumina, including wear-resistant and special structural high-strength ceramic bodies.

Perhaps the most important application for this family of ceramic bodies has been in electronic packaging. One of the first used in this area was for the now famous ceramic-dual-in-line packages (Cer-DIPs). These range from ~4 to ≥20 mm in width and length dimensions and on the order of 1–2 mm in thickness. The Cer-DIPs usually have recesses into which semiconductor chips are attached. The chips are wire bonded using gold wire, which is attached to metal leads. Special glasses are used to insulate and attach covers of the same Cer-DIP composition, rendering hermetically sealed packages. There are many configurations of this technology, including those in which leads can be present on four sides of a "Quad" package. Thin layers of the green ceramic body can be doctor-blade cast as tape and then printed with conductive patterns before being stacked and fired to make a multilayered package, etc.

Although plastics are becoming increasingly common as a package material, the original Cer-DIP composition, with small, proprietary modifications, remains the material of choice for packages for most semiconductor chips. Stability, strength, opacity, compatibility, and ease of manufacture make this almost a universal choice for electronic packaging.

Literally millions of Cer-DIP or related pieces using this basic formulation are made and put to use worldwide, every day. Few, if any, electronic communication devices, including radio, television, satellites, space vehicles, and computers, do not contain one of more of the packages made possible with this technology. A common saying is that, "This [technology] made it possible for the, often tiny, chips to be brought out into the real world where they could be used."

The developers recognized the uniqueness of this innovation, while perhaps not fully appreciating its limits. This work resulted in a series of papers, the most important of which is "Sintering of Alumina at Temperatures of 1400°C and Below," by Ivan B. Cutler, Cyril Bradshaw, Carl J. Christensen, and Edmond P. Hyatt, published in the *Journal of the American Ceramic Society*, **40** [4] 134–39 (1957). This author has found current Russian and Asian alumina manufactures referencing this original paper. U.S., European, and Japanese ceramic package manufacturers all use this innovative concept, which was announced 40 years ago.

Pore-Free Ceramics

by Joseph Burke, General Electric Company, retired

During and following World War II, a variety of inorganic, nonmetallic materials became technologically important because of their special properties. Fabricated from powders, they were consolidated by heating. This process, called sintering, differed from classical silicate ceramic processing in that no liquid phase was formed at the high temperature to assist matter transport. (Now the term sintering includes both solid-state sintering and liquid-phase-assisted sintering.)

All such solid-state-sintered ceramics were porous after firing, and, hence, opaque, because light is diverted at the solid–vapor interfaces. Studies to determine the mechanism by which solid-state-sintering consolidation occurs became an important part of the field of materials science that was then developing, and elegant work by Kuczynski and by Kingery and Berg, among others, demonstrated that, contrary to prior opinion, solid-state diffusion was the important mechanism of transport and that specimen shrinkage can be considered to involve the "evaporation" of lattice vacancies from pores and their subsequent annihilation at grain boundaries. Simply, this implies that, with sufficient heating time, all pores in a specimen should be annihilated. However, Alexander and Baluffi showed that, in copper specimens, pores are eliminated in reasonable sintering times only when they are intersected by grain boundaries. Burke subsequently showed that, although all pores in such particle aggregates initially lie on grain boundaries, during sintering heat treatment, they are usually entrapped inside grains by a special grain-growth process called (by metallurgists) discontinuous grain growth, and, if so trapped, they are substantially permanent. He also observed that a particular alumina specimen with a (then) unidentified impurity did not undergo discontinuous grain growth; therefore, pores remained on grain boundaries even when the specimen was greatly over fired, and it did sinter to a substantially pore-free state.

Subsequently, Coble, working with Burke, found that a small amount of magnesium oxide would inhibit discontinuous grain growth in aluminum oxide and permit it to be sintered to theoretical density, to yield a translucent product with excellent light-transmitting properties. Now translucency was available in at least one of the new ceramics, and the new property led to important new products.

The first product, called Lucalox Alumina, has found use primarily as the envelope for a discharge lamp using high-pressure sodium vapor, which operate at a high temperature. It has a luminous efficacy of well over 100 lumens/W, the highest of any light source (a 100 W tungsten filament lamp has an efficacy of ~18 lumens/W). It has an almost continuous spectrum, which is deficient in the blue-green and, hence, has a golden color. It is now made by many manufacturers, and, because of its high efficiency, it has a large part of the outdoor-lighting market. A new product, the ceramic–metal halide lamp, that utilizes the same ceramic envelope, has an intense white light and is just now being introduced.

In general, the new understanding taught that powders can be sintered in the solid state without externally applied pressure to a pore-free state but that competing processes, such as discontinuous grain growth, can effectively prevent the elimination of pores. Utilizing this and other new understanding to suggest suitable additions and proper processing controls, other important pore-free ceramics have been developed. Pore-free silicon carbide, by Svante Prochazka, and Scintillators for medical X-ray detectors, by Charles D. Greskovich, are described in these pages.

Pore-Free Silicon Carbide Ceramics

by Svante Prochazka, General Electric Company, consultant

Silicon carbide and silicon nitride were considered in the 1960s as the best candidates for heat engines with increased operating temperatures and higher performance. Because these substances do not melt and their powder compacts do not densify on sintering, processes such as hot pressing and reaction bonding were applied to consolidate them. These fabrication processes were not satisfactory for material for gas turbines and other heat engines. Reaction sintering compromised some of the properties while hot pressing was too expensive and severely limited shape capability.

A sintering technique is desirable, but silicon carbide did not respond to sintering attempts. When its powder compacts were heated to high temperatures, no shrinkage occurred; when heated above 2300°C silicon evaporated and graphite was left behind. This was really not surprising. Silicon carbide is a strongly covalently bonded solid, and covalent substances, such as carbon, boron, silicon, and germanium were known not to sinter to dense bodies. It is the strong interatomic bonding that gives silicon carbide the potential to work under extreme conditions, but is also the reason why it is such an intractable substance. In fact, exceedingly high pressures were necessary to consolidate pure silicon carbide powders by hot pressing into dense bodies.

R. A. Alliegro in 1958 invented the use of small additions of aluminum, boron, iron, or other elements in hot-pressing silicon carbide to lower the necessary pressure and temperature. In 1973 S. Prochazka observed that pressure could be further reduced by using a carbon-rich silicon carbide powder with boron additions. Moreover, using very fine silicon carbide powder with these additives made it possible to densify compacts even without pressure, just by heating them in an argon atmosphere at 2100°C. Thus, in the late 1970s a versatile process for making inexpensive, nonporous silicon carbide parts became available in which one could apply all the routine ceramic-shaping procedures, such as isostatic pressing, green machining, slip casting, extrusion, tape casting, and injection molding. The sintered material is composed of grains of silicon carbide a few micrometers in size and a certain amount of boron carbide and graphite grains depending on the amount of additions. The two latter phases can be varied from practically nil to a substantial fraction, which makes it possible to modify some properties of the product.

The original dream, to build a gas turbine with a silicon carbide hot section has not yet materialized and probably never will. The thermal shock resistance and fracture toughness proved insufficient and the strength marginal for such an application. However, its hardness, wear resistance, hot strength, electrical and thermal conductivity, high creep, and chemical resistance made sintered silicon carbide an attractive alternative where other ceramics were failing or proved insufficient. Seal rings for pumps and diesel engines, sand-blasting nozzles, tubes for roller hearth furnaces, heaters, igniters, sheathing for sensors, and many custom-made parts are listed among the products manufactured of sintered silicon carbide by more than a dozen companies.

Special Chemical Processing

Early History of Sol–Gel Ceramics

by Thomas E. Wood, 3M Company

Although new applications for this technology continue to emerge, sol–gel technology is characterized by a rich and extensive history. The scientific principles of sol–gel, i.e., the use of gels as metal oxide precursors, gelation of a colloid to form a ceramic green piece, the use of soluble ceramic precursors, and the mixing of soluble precursors to form complex oxides, were developed long ago. For example, in the early 18th century, chemists discovered that metal hydroxides and gels could be used as precursors to metal oxides. In 1779 T. Bergman discovered that, if a proper measure of a dilute acid were added to a water-glass solution, the whole mixture could be converted into a gel. Soluble silica or water-glass quickly found extensive use as a binder in the manufacture of fire brick and porcelain. By the early 1900s, coprecipitation of metal hydroxides was used to synthesize ferrites. Thus, by the beginning of the 20th century, the foundation for sol–gel ceramics had been laid.

One of the earliest industrial applications of sol–gel chemistry, first described by Kessler in 1883 and still in use today, involves the treatment of calcareous stones with salts of fluorosilicic acid. In this process, the fluorosilicic acid, upon contact with the basic carbonate stone, undergoes hydrolysis to yield exceedingly fine precipitates of magnesium and calcium fluoride in a matrix of amorphous silica. This process remains the treatment of choice for care of marble floors in most of southern Europe.

In the early 1900s one of the fathers of modern sol–gel chemistry, W. A. Patrick of Silica Gel Corporation, began his pioneering work in the area of sol–gel-derived desiccants, adsorbents, and catalysts. In 1918 he reported that very porous silica could be prepared by carefully drying and firing a homogeneous, salt-free silica gel. After firing to temperatures of up to 700°C, he described the product as "a hard, transparent substance resembling glass very closely in appearance." He found the transparency to be dependent on both water content and pore size. In 1923 he prepared silica-containing catalysts by impregnating partially dried silica gels with solutions containing $(NH_4)_2PtC_{16}$ or $Fe(OAc)_3$ followed by controlled drying and firing. In 1926 he extended his technology to include gel-derived oxides of iron, chromium, manganese, copper, bismuth, lead, thorium, nickel, and vanadium. In 1927 Patrick patented the process to prepare mixed gels containing silica, and in 1928 he patented sol–gel methods to prepare porous tungsten oxide, stannic oxide, aluminum oxide, and titanium oxide. The work of Patrick served to establish what is now known as sol–gel technology as an industrial process to synthesize adsorbents and catalysts. Interestingly, W. A. Patrick Jr. also later developed gel-derived oxides as adsorbents.

The application of metal alkoxides in ceramics and sol–gel technology began with the use of these materials in the preservation of stone and rapidly spread to the generation of ceramic coatings. In 1927 King and Threlfall patented the use of partially hdyrolyzed alkoxysilane solutions to generate coatings and to impregnate porous bodies. In 1931 an article by King, "Silicon Ester Binder," outlined methods for hydrolyzing alkoxysilanes for application in paints, dental cements, stone preservation, and treatment of brick and concrete. In a 1934 patent A. B. Ray taught the deposition of silica via solutions of hydrolyzed alkoxysilanes. C. P. Marsden developed an industrial method of coating silica on the interior walls of light bulbs and vacuum bulbs using hydrolyzed alkoxysilane solutions in 1938. In that same year, Dan McLachlan of Corning Glass Works discovered that finely divided oxides could be prepared by hydrolysis of alkoxides or chlorides by an emulsion technique using an inert medium. In 1939 Walter Geffcken and Edwin Berger patented a novel and effective treatment of glass surfaces that involved spraying a precursor fluid on a heated glass substrate to precipitate a gel-like, hydrated oxide on the glass surface. By about 1953 this technology was being used commercially to apply antireflection coatings on automobile mirrors. The pioneering work of Geffcken was followed up by Hubert Schroder and Helmut Dislich at Schott Glaswerke in the early 1950s. In 1941 Carl Christensen of Bell Telephone Laboratories showed that organic–inorganic composites could be prepared by hydrolysis of alkoxysilanes in the presence of organic fillers. Soviet scientists, such as Kreshkov, disclosed that hydrolysis of alkoxysilanes containing finely divided oxides and hydroxides could be used as a route to ceramic materials. By 1946 the use of silicon alkoxides was well established in industry, as shown by a noteworthy article by Cogan and Setterstrom that described the properties of tetraethoxysilane and its use as a source of silica.

Some of the greatest contributions to the development of early sol–gel technology were made by industrial researchers in the areas of catalysts and catalyst supports. The early 1940s witnessed dramatic advancements in the application of sol–gel technology to these areas. A quote from a 1944 British patent conveys somewhat the flexibility of the sol–gel process as utilized by the catalyst industry, "As a generally suitable method of preparing the fluid catalysts, at least one major constituent of the catalyst is prepared as a sol or gel or gelatinous precipitate by proper mixing of the necessary ingredients. Other components when employed, are preferably added as sols or gels, which may be formed concomitantly with the first sol or gel, mixed with it later, or formed on it, or they may be incorporated as non-colloidal precipitates or as ions whether adsorbed on the gel or dissolved in the dispersing medium .... In some cases a change from gel to sol or sol to gel may be made by the addition of a suitable peptizing or coagulating agent."

Beginning in the 1960s the sol–gel approach began to be successfully applied to the preparation of thoria and urania microspheres for the nuclear industry. For these nuclear applications, new gelation chemistries were established and the sol–gel processes were shown also to be useful in preparing non-oxides. It is possible that the term sol–gel was coined by workers in the nuclear industry to describe this technology. However, early workers in the nuclear field recognized that this technology had an earlier origin as it was applied much earlier in the catalyst field.

At present, sol–gel technologies continue to attract considerable attention as a family of versatile and powerful synthetic methods. Besides the fields of catalysts, adsorbents, ceramic coatings, and the nuclear fuel area, sol–gel has continued to find new applications in areas such as ceramic filters, ceramic fibers, specialty glass materials, abrasive minerals, and biomaterials. A consideration of the history of sol–gel reveals that this technology reflects the culmination of the contributions of numerous, diligent scientists working at their art for well over a century. With continued emphasis on this area, sol–gel technologies will certainly prove to be relevant well into the next millennia.

Sol–Gel Processing of Ceramics

by Rustum Roy, The Pennsylvania State University

The traditional method of making ceramics since time immemorial was to mix thoroughly solid ceramic powders, shape the mixed powders, and fire the shaped body. In the laboratory it was precisely the same. In 1947–1948 Rustum Roy, working as a student (then a postdoc with Professor E. F. Osborn), realized that such mixing, especially for getting equilibrium in the system Al_2O_3–SiO_2 and Al_2O_3–SiO_2–H_2O, was out of the question. Out of that frustration was born the sol–gel process.

Roy clearly saw that the only way to get molecular-scale mixing was to put all the ions in solution and mix them in solution. Once in solution the next step was to coprecipitate them (as was done in the system MgO–Al_2O_3), or to form a gel to "freeze-in" all the atomically mixed ions. Drying the gel and decomposing the organic or water complexes at 500+°C gave ceramic mixtures both extremely fine and extremely pure as the standard starting powders for high-science research in ceramics. That is precisely where the sol–gel process has come to after 50 years—it is the preferred universal general process for making pure or high-tech or very reactive ceramics. At some American Ceramic Society meetings, almost one-third of the papers relate to sol–gel science.

Roy originated the use of organic precursors right from the beginning, in 1948–1949—sometimes synthesizing his own organometallics. By 1956 he and his students had made several thousand ceramic compositions—which exist today—in binary, ternary, and quaternary systems involving SiO_2, Al_2O_3, MgO, CaO, FeO, K_2O, Na_2O, BaO, TiO_2, ZrO_2, Ga_2O_3–GeO_2, NiO, etc. Toward the end of this period, it was found that it was much cheaper and easier to use the commercial sols of oxide—such as Ludox, which were coming on the market—instead of the organic precursors.

One remarkable commentary on the innovative process is the fact that, for 10–15 years, only Pennsylvania State University students and postdocs used the process. The geoscience community did adopt it in part, but largely through Pennsylvania State University students and visiting faculty.

In industry, Dislich, in the Schott Company, was using a variant of the process using SiO_2 and TiO_2 sols for making coatings on glass windows at the end of World War II. As a result, coated windows became a big industry and the sol–gel process had its first exploitation in industry. At the Oak Ridge National Laboratory, an ingenious adaptation of getting the oxide solution gelled by allowing drops to sink through an organic liquid base (hexamethylene tetramine) resulted in nuclear fuel pellet beads, another first ceramic product via sol–gel. A major use resulted from 3M using a sol–gel route to abrasive grain manufacture. Trial pots of optical glass were made. Owens-Illinois attempted a full tank to test the economics. In the laboratory and the capacitor industry, and now in superconductor substrate solution, sol–gel route has become the standard for much of their film work, and for much routine mixing for general ceramic research.

The importance of this development, especially in ceramic science, can be gauged by the fact that Roy's final review paper in the *Journal of the American Ceramic Society* in 1956 became the first citation classic ever, in the field of ceramics, and the first such ever published in the *Journal.*

Sol–Gel Ceramic Products

by Harold G. Sowman, 3M Company, retired

Sol–gel technology, sometimes referred to as chemical ceramics, is a term applied broadly to processes for the preparation of ceramics based on the shaping of liquid or plastic precursors that are gelled or polymerized. The shaped precursors are fired in a controlled manner to remove fugitives and sinter the inorganic phase. Broadly speaking, sol–gel technology utilizes sols, solutions, and mixtures as sources of the ceramic articles. Sols generally comprise fine particles that are <100 nm in size in liquid suspension. Such sols may be transparent or translucent. Particles are much smaller than those in the more traditional slips or slurries.

The commercial availability of many sols and metal-containing chemicals, especially alkoxides, has contributed to the viability of the technology. In some cases, the chemical compound solution provides properties necessary for shaping and rigidizing whether used alone or mixed with sols. A review of the patented art shows that many of the ceramic articles are prepared from mixtures of sols and solutions. Forming properties of sol systems can be enhanced, in some cases, by the addition of completely fugitive organics, such as poly(vinyl pyrollidone), poly(vinyl alcohol), or even corn syrup.

The molecular or near-molecular scale of the inorganic components of sol–gel processing makes possible the control of early stages of crystallization and manipulation of combinations of materials so as to provide microstructures, crystalline species, surfaces, and other properties not attainable with other processes. Transparent microcrystalline ceramics, such as fibers, microspheres, and coatings, that look similar to glass can be made. Melting is not required, and sintering temperatures are relatively low.

Ceramic fibers can be prepared by spinning or extruding and drawing viscoelastic mixtures of metal-containing sols and solutions. Fiberizing and firing can be a continuous process wherein multifilament strands are wound continuously on spools. Beads or microspheres are generally formed by dispersion of droplets of the sol precursor in liquid media that may be immiscible or partially miscible with the liquid of the sol system. Granules are products of dried and fractured gelled mixtures that are sized and fired to yield ceramic granules for abrasives or other applications.

Many industrial companies have become involved in sol–gel technology. A broad-based program at 3M Company on sol–gel or chemical ceramic technology initiated by Harold G. Sowman in the 1960s has resulted in many commercial products. Some of these products include ceramic fibers, abrasives, transparent reflective ceramic beads for highway marking, protective ceramic coatings of phosphor particles for electroluminescent lighting systems, and ceramic fillers for dental resins and cements. Ceramic fibers can be stiffer and endure higher temperatures than glass fibers and are used for reinforcement of metals and ceramics as well as resin and for thermal protection and insulation in fiber or fabric form (e.g., thermal protection system for the space shuttle) and in high temperature filtration. Abrasive granules manufactured via sol–gel processes last several times longer than conventional abrasive minerals and highway marking strips using reflective ceramic beads are many times more durable than those using glass beads.

In addition to 3M, ceramic fibers have been manufactured by Nippon Carbon Company, Ltd., E. I. du Pont de Nemours and Company, Inc., Imperial Chemical Industries, plc Ltd., and Sumitomo Chemical Company, Ltd. The Norton Company, also is known as a manufacturer of abrasives with this technology.

Large Silica Glass Bodies from Colloidal Sols

by David W. Johnson Jr. and John B. MacChesney, Bell Laboratories, Lucent Technologies

Sol–gel technology has engendered widespread research resulting in thousands of published papers. The process has the advantages of extreme chemical homogeneity and net-shape forming along with small particle sizes leading to low-temperature sintering. It has demonstrated, both for colloidal dispersions of nanoparticles and for hydrolysis and condensation of metallorganic compounds, formation of powders, films, and monolithic bodies.

Despite the research effort, sol–gel processing has not been well developed for producing large monolithic bodies. Impediments have largely centered around the large shrinkage that must take place in drying gels where the capillary stresses endured in fragile materials lead either to cracking or excessively slow drying rates. Bell Laboratories has developed a colloidal sol–gel process that has been successful in producing bodies of SiO_2 glass >4 kg while taking advantage of its net-shape processing and high purity.

Optical fibers are produced by forming a glass preform consisting of a core glass of higher index of refraction surrounded by cladding glass of lower index of refraction. This preform is then drawn into optical fiber. The core glass and immediate cladding can be deposited in a tube using modified chemical vapor deposition (MCVD) or by flame hydrolysis of precursors ("soot" methods). In either case the core is typically SiO_2 glass doped with GeO_2 to increase the index, and the cladding is SiO_2 glass. It is economic to fabricate large preforms. One means of attaining large preforms is to shrink (on a glass working lathe) a large SiO_2 glass tube over a core rod that contains the optically active core and immediate overcladding. Tubing glass of the purity and dimensional quality needed for overcladding applications is expensive. This is one application for sol–gel SiO_2 glass. This process begins with fumed silica that is dispersed in water to a concentration of 46 wt%. An organic base, such as tetramethylammonium hydroxide (TMAH) is added to bring the sol to a pH of 11–12. The sol is centrifuged to remove large impurity particles and then deaerated using vacuum. An ester, such as methyl formate (MF), is added, which slowly reacts with water to produce formic acid and reduces the pH of the sol. The sol is cast into a precision mold consisting of a polished metal cylinder with a concentric core rod. After 5–15 min, the sol reaches a pH of ~9, where it gels. After an equilibration period, the center mandrel is removed and the gel body is pushed into a water bath. From there it is placed in a drying chamber, where it is rotated to maintain its shape in an atmosphere of controlled temperature and humidity.

After it is dried, the body is heat-treated in two steps. The first is to remove remaining water and organics by heating to 600°C. Heat treatments in Cl_2- and $SOCl_2$-containing atmospheres in the 600°–1000°C range remove hydroxyls, transition metal, impurity ions, and refractory oxide particles, such as ZrO_2, which would cause low-strength breaks in the fiber. Finally, the part is sintered to transparent glass by passing it through a 1500°C hot zone in a helium atmosphere.

Such bodies are >1 m long weighing >4 kg. Dimensional precision has been demonstrated to within 0.7%, and the tubes are straight to within 0.5 mm. Concentricity and ovality also meet the standards necessary for commercial applications. When shrunk into a core rod and pulled into fiber, the fiber exhibits losses on the order of 0.33 dB/km at 1310 nm and <1 internal break per 1000 km of fiber.

In this demonstrated application, the glass is not optically active, but the purity is such that it is possible to utilize this glass much closer to the core. Neutron activation analysis shows this glass to have impurity levels of <10 ppb for most ions. Also, the ability to form net-shape articles makes it appealing for other applications of vitreous silica.

3

Basic Glass Processing

For the purposes of this book glass products have been grouped into flat glass, bottle-shaped items, fiberglass, and glass specialty items.

A major innovation in flat glass production has swept the world—the Pilkington float glass process. Another major innovation, the ribbon machine for glass bulbs is also in worldwide use. Other important innovations in basic glass processing include the suction bottle machine, the process for making continuous glass tubing, and the continuous melting of optical glass.

Glass fibers have been known from ancient times and pilot production was attempted at the turn of the century. However, the real commercial breakthrough came in the 1930s with the mass production of both continuous glass fibers and mats of blown fibers.

Specialty items treated in this section include glass-ceramics used in radiant glass-ceramic cooktops, photochromic and photosensitive glasses, glass microspheres, laminated glass, and borosilicate laboratory and consumer glassware. Glass TV tubes have swept the world and appear to be taking another step to large, flat faces.

Glass Windows, Bottles, and Bulbs

Float Glass Process

by William R. Prindle, Corning Incorporated, retired

Flat, distortion-free glass has long been valued for major windows, doors, mirrors, display cases, etc. For centuries plate glass was cast, rolled, ground, and polished to produce an attractive, high-quality product. The process, however, was labor intensive, required much handling of the glass, and had high waste-glass losses. Accordingly, plate glass was expensive and a premium product. Drawing processes were used extensively for window glass, but were not suitable for producing a distortion-free product for the more demanding applications.

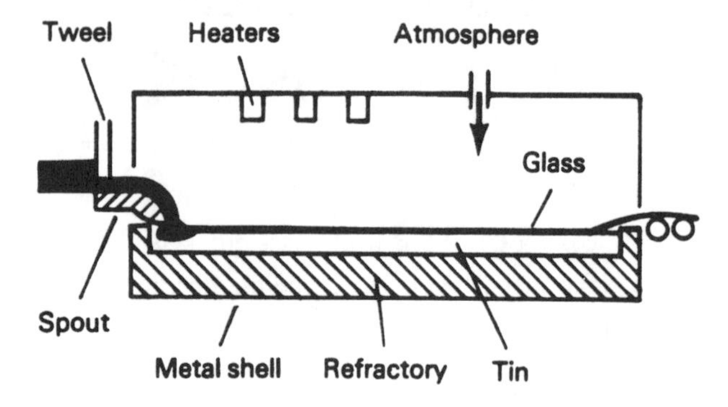

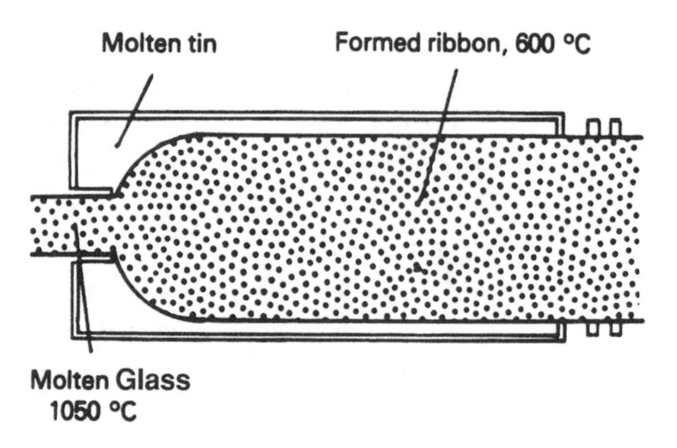

Figure 8. *Side and top schematic views of the float glass process. Courtesy ASM International, from* Engineered Materials Handbook, *Vol 4,* Ceramics and Glasses *(1992), ASM International, Materials Park, OH 44073-0002, p 399 (Fig. 6).*

In the late 1950s the float glass process was introduced to make large unblemished glass sheets at a reasonable cost. In the float-glass process a stream of glass flows out of the melting furnace on to a bath of molten tin. The layer of glass is maintained at a temperature high enough for it to spread out and for its top and bottom surfaces to become flat and parallel. The molten tin surface is flat and the forces of surface tension plus gravity impart this flatness to the bottom of the sheet. The top surface, kept hot by the heated atmosphere of the bath chamber, is also flattened by these forces. The broad glass ribbon is cooled gradually as it is drawn along the surface of the tin bath until it is rigid enough to be lifted from the tin and introduced to the annealing lehr without marking the glass. The emerging glass has fire-polished surfaces, is of uniform thickness, and requires no further grinding or polishing.

The float-glass furnaces are among the largest glassmelting tank furnaces in use today and can produce 800–1000 tons of finished glass per day. A float glass production line (tank, tin bath, and lehr) can be 700 ft long with a tin bath >150 ft in length and wide enough to produce a sheet with a width of 12 ft. A slightly reducing atmosphere is maintained over the tin bath to inhibit oxidation of the bath.

Alastair Pilkington was the innovator who conceived the process in 1952; in 1959 the process was announced and sales of the new product were made on a commercial scale. The float glass process soon displaced the traditional grind-and-polish plate glass process, and, today, virtually all "plate" glass is made in this manner throughout the world. Many billions of dollars worth of glass are produced by this process every year.

Enhanced Float-Glass Process

by Tsuneharu Nishikori, Asahi Glass Company

The normal float-glass process developed by Pilkington can produce plate glass with completely parallel surfaces of 2–25 mm in thickness. It has greatly contributed to the building, mirror, and automobile industries.

Asahi Glass Company has added entirely new technologies to the float-glass process in the course of an effort to meet market demands. These are the technologies for forming of (1) very thin (<1 mm) glass and (2) glasses with compositions other than the ordinary soda–lime glass.

The very thin glass was strongly requested from Japanese manufacturers who were leading the world LCD (liquid crystal display) market. Several production processes other than the float-glass process were proposed. Recognizing the principle of the float glass process that the surface of a liquid is completely flat stimulated a challenge by Yoshiaki Kondo to extend its use to thicknesses <1 mm. A team headed by Hajime Amemiya finally succeeded. As a result, Asahi Glass Company began supplying very thin float glasses in 1982 for the first time in the world.

The enhanced float-glass process uses several key technologies: to stretch glass ribbon over the tin bath to a paper thickness; to achieve surface flatness even in the region of paper thickness where the situation deviated far from the equilibrium thickness of 6 mm; to handle (i.e., cut, separate, convey, and pack) extremely thin and wide (2–3 m) glass ribbon running at high speed (several hundreds of m/h) without scratching its surfaces; to simulate and evaluate the new processes using computer calculations; to design and manufacture the equipment; and to construct the total process.

Asahi's next challenge was to apply the float process to new glasses, such as alkali-free silicate, aluminosilicate, and borosilicate glasses, that had never been attempted in the float process. These glasses are now expanding their roles and are expected to play more leading roles in the future. Among other difficulties, overcoming high viscosities had to be accomplished. The working temperatures of these glasses are 100°–200°C higher than that of ordinary soda–lime glass. The original float process could not cope with these high temperatures. Tetsu Mori and Kisuke Tajima succeeded in developing a new float-glass process operable at such high temperatures. The new glasses produced by the new process were defect-free and had excellent flatness. Substrates for the TFT-LCD with 0.5–0.7 mm thickness and for the PDP (plasma display panel) with 3 mm thickness fulfill the customers requirements for surface smoothness, light weight, strength, chemical stability, electrical insulation, etc.

The large-area capability of the float-glass process is an excellent advantage over other production processes, and the excellent technology already applied to this process for a variety of glass compositions to avoid the use of arsenic, a very convenient refining agent, will become ecologically more important for generations to come.

Ribbon Machine for Glass Bulbs

by William R. Prindle, Corning Incorporated, Retired

Thomas Edison's invention of the incandescent electric lamp in 1879 created the need for a reliable supply of sturdy, low-cost glass bulbs. Although this need was met initially by glassblowers using conventional off-hand blowpipe technology, it was soon followed by blowing the bulbs in paste molds, wherein the glassworker rotated the hot glass bulb in a mold lined with a wet charcoal coating as he blew. The cushion of steam formed gave the bulb a smooth surface and yielded a product with spherical symmetry. This technique was used at the turn of the century with teams of four glassblowers turning out more than 1200 bulbs a day.

The swift spread of electric lighting soon required a faster, more efficient bulb-manufacturing process. The first devices designed to mechanize the process simulated the motions of the human glassblowers, and various machines appeared in the early decades of the 20th century that produced up to 40 bulbs a minute.

William J. Woods, a young production engineer at Corning Glass Works, thought of a concept for a totally different type of machine in the early 1920s, and after a few crude demonstrations was assigned to work with Corning Chief Engineer David E. Grey to design and build a prototype machine. The new device, known as the Corning bulb machine, or ribbon machine, was put into production in 1926 and was soon producing bulbs a the rate of 250 per minute.

The ribbon machine receives a uniform stream of molten glass ~3 cm in diameter from an opening in a glassmelting tank furnace. The glass is pressed between two water-cooled metal rolls, one smooth and the other with cuplike depressions at regular intervals in the roll face. The stream of glass is converted by the rolls into a flat ribbon that has patties or buttons of molten glass on it at the spacing set by the roll depressions. The ribbon is then received on a moving endless belt with each button accurately positioned over an opening in the belt. The molten glass buttons sag through these openings to form hollow pear-shaped drops, similar in shape to the final bulbs. As these soft bulbs gradually increase in length while the belt moves, paste molds with wet charcoal linings, moving at the same speed as the belt, close about them from below, and blow-heads moving at the same speed close down tightly on the ribbon from above. The molds rotate rapidly about the bulb to ensure a uniform steam cushion as the air blown from above increases. The belt carries its ribbon of glass forward continuously, with each button having been blown into a fully formed bulb. The molds then split apart and fall away, leaving the finished bulbs suspended from the ribbon. The bulbs are separated from the ribbon by a rotating knife that gives each a light tap and are caught on an annealing belt.

The ribbon machine has gone through many engineering improvements in the three-quarters of a century since its invention, increasing its output to a many as 200 40-W bulbs per minute, although the principle and basic machine remain unchanged. One machine, operating three shifts a day, can produce more than 2,000,000 bulbs per day, even with some down-time. Today, almost all the light bulbs in the world are made by a relatively small number of ribbon machines (a few special shapes or sizes are made by other means), and inexpensive electric lighting has brightened the lives of millions.

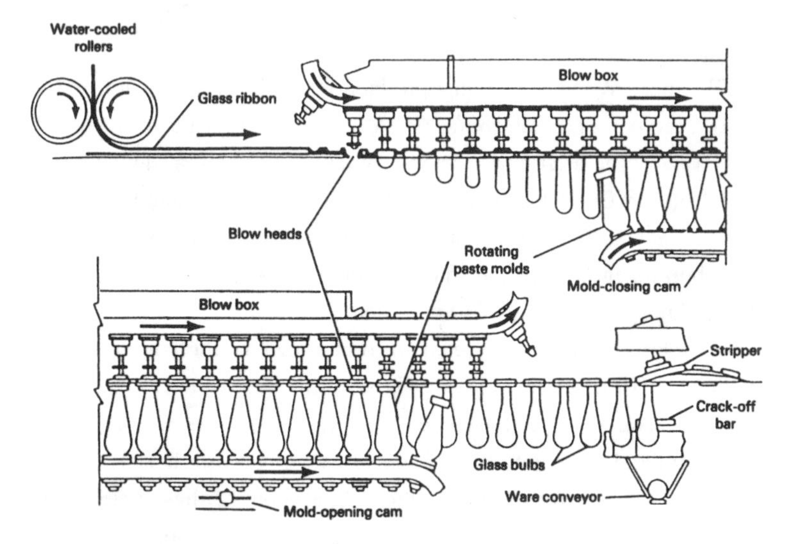

Figure 9. *Schematic of the ribbon machine for making glass bulbs. Courtesy W. R. Prindle.*

Ribbon Machine for Glass Light Bulb Forming

by William Rhodes and Robert Alspaugh, Osram Sylvania, Inc.

Before the invention of the ribbon machine, light bulbs were blown by hand. One man could make ~150 per day. Edison's search for a company that could make bulbs quickly and cheaply brought him to Corning Glass Works and ultimately the Osram Sylvania, Wellsboro, PA, glass plant, which was then a part of Corning Glass Works. It was there that William Woods, a Corning engineer, who, as coincidence would have it, was born the same year as the light bulb, developed the ribbon machine.

Woods' idea was a machine that substitutes compressed air for human breath in blowing the bulbs. As a ribbon of molten soda-lime glass the width of a man's hand travels down the machine, air-blowing plungers descend upon the glass and blow it into a mold. At the end of the machine, the bulbs are knocked onto a cushioned conveyer belt and carried to packing lines.

The machine's modernization allows this to happen at incredible speeds. Today, the ribbon machines at Wellsboro produce anywhere from 1200 bulbs per minute (standard A-19, 60 or 100 W bulbs) to 2000 bulbs per minute (small automotive and specialty lighting products). The machine uses modern materials, such as Kevlar, Teflon, and Inconel, and uses sophisticated temperature control and electronic control processes to increase the efficiency. It also uses a computerized vision system, which identifies defective bulbs before they are packed and shipped to the customer. Although the ribbon machine has been computerized and otherwise modernized, the underlying technology—blowing compressed air onto a ribbon of molten glass, creating the lamp envelope—has not been changed. The technology is the standard used all over the world in making light bulbs.

The American Society of Mechanical Engineers designated the ribbon machine the 10th International Historic Mechanical Engineering Landmark, an honor it bestows on inventions that have changed history.

Owens Suction Bottle Machine

by William R. Prindle, Corning, Incorporated, retired

The development of the automatic suction bottle-forming machine by Michael J. Owens in 1903–1904 was probably the most significant improvement in the manufacture of glass containers since the invention of the blow-pipe about two millennia earlier. The use of glass containers increased rapidly in the 19th century because of both the use of pasteurization and hermetically sealed food and drink packages, and to the general population growth. To meet the increased demand, by the end of the century, there were many semiautomatic machines in operation that depended on the use of multiple blow-pipes, each with a hand-gathered gob of glass at the tip.

Owens, an experienced glass worker and plant superintendent for the Libbey Glass Co. of Toledo, OH, worked for about 10 years on various approaches to mechanize the bottlemaking process, focusing in particular on devices that gathered the molten glass by suction. Owens was supported financially be Edward Drummond Libbey, and they formed jointly first the Toledo Glass Company and later the Owens Bottle Machine Company to exploit the rapid success that Owens achieved in machine development.

After a short series of design changes, an automatic rotary machine was built that carried six arms, each equipped with a gathering mold and a blow mold. As the machine rotated, each arm successively was lowered so that the gathering mold dipped lightly into a small rotating pot or bath of molten glass, and suction was applied to fill the mold. As the filled gathering mold was lifted from the glass surface, a cut-off knife sheared off the hot glass cleanly at the bottom of the mold. As the machine rotated further, the gathering mold opened and dropped away, leaving the hot glass held by a neck ring that formed the bottle opening. A blow mold then closed around the gather, and the machine blew the bottle to its final shape; a bit more rotation and the finished bottle was taken out mechanically and conveyed to the annealing lehr.

The rotating pot from which the glass was gathered rotated in the opposite direction from the bottle machine so that each gathering mold encountered unchilled glass of a uniform temperature as the suction was applied. This small rotating pot was fed continuously from a large glass tank melting furnace.

Owens made continuous improvements to the machine, doubling the output per machine between 1903–1904 from 100 gross of pint beer bottles per day to 200 gross per day by 1909. He then added extra arms and molds to the machine, first building a 10-arm machine in 1912 and then a 15-arm machine in 1914, the latter capable of making 500 gross of pint beer bottles per day. The direct costs dropped from $1.50 per gross in 1903, using conventional blowing practices, to $0.1 per gross by 1907, using the Owens machine.

The Owens suction machine became the bottle-manufacturing process of choice and dominated the industry from ~1905 until the late 1930s, when it was superseded by various gob-fed automatic machines, such as the Hartford-Empire individual section (IS) machine. There are no Owens suction machines known to be operating in the United States today. The Owens name, however, has been preserved in the names of major U.S. glass companies, such as Owens-Illinois, Owens-Brockway, Owens-Corning Fiberglas, and Libbey-Owens-Ford.

Danner Process for Making Glass Tubing

by William R. Prindle, Corning Incorporated, retired

Glass tubing is used in substantial quantities in electrical illumination (e.g., fluorescent and neon lighting), chemical apparatus, and medical products, and until 1917 was all drawn manually. The classic procedure, in use for several centuries, started with a gather of molten glass blown and worked into a short, hollow cylinder with thick walls. An iron rod or punty was attached to the opposite end from the blow-pipe, and then the cylinder of hot glass was quickly stretched out to form a narrow tube. This was accomplished by each worker walking away from the other as they held the blow-pipe and punty. At any point where the diameter of the tube showed a tendency to become too small, a third worker cooled the glass to stiffen it by fanning it or by blowing air on it to prevent further narrowing. This crude process made it very difficult to maintain a uniform bore in the tubes and also led to a high rate of breaking from the unequal cooling; accordingly, only ~25% of the glass was usable.

In 1917 Edward Danner of the Libbey Glass Company invented an automatic method for making glass tubing. In Danner's process, molten glass of the appropriate viscosity flowed continuously from a spout in a melting furnace down on to a slowly rotating (~8 rpm) refractory mandrel or blow-pipe that was inclined at ~15° down from the horizontal. The glass flowed together to form a continuous coating on the mandrel, and the stream of air passing through the mandrel kept open a bore of constant diameter in the tubing that was formed. This tubing was then pulled mechanically at a constant rate to rolls that supported it as it cooled. Glass temperature, air pressure, and drawing speed could be adjusted to adjust the size of the tubing.

The Danner process reduced the total cost of making a pound of tubing by two-thirds and the direct labor costs by 80%–85%. The process was quickly accepted and was licensed by Libbey to many other companies; by 1926 hand-drawn tubing had all but disappeared.

The Danner process continues in use, although other methods for making tubing, such as the down-draw Vello process, were introduced subsequently. All the processes, however, share Danner's concept of a continuous supply of molten glass, a consistent stream of low-pressure air inside the tubing, and a steady drawing mechanism to keep the bore diameter constant.

Continuous Melting of Optical Glass

by William R. Prindle, Corning Incorporated, retired

Although glass for containers and flat glass had been melted in continuous tank furnaces since the latter part of the 19th century, the continuous melting of optical glass dates from the early 1940s. The applications of optical glass for prisms and lenses in optical instruments require a level of homogeneity far higher than that acceptable for bottles and windows, with the most demanding requirement being freedom from striae or fine cords. This imperfection is an attenuated glassy threadlike inclusion differing slightly in composition from the bulk glass, and according, differing slightly in refractive index and other properties. There are many causes for striae, but in optical glass they are typically the result of solution of refractories in the base glass, and incomplete blending of the various glass consituents as they melt.

The invention by Guinand around 1790 of stirring pots of molten glass with a refractory ceramic stirrer was one of the greatest contributions to improving the homogeneity of optical glass, but the process continued to produce glass with excessive striae much of the time. Michael Faraday, in the late 1820s, demonstrated that melting glass in platinum also improved homogeneity, but his experiments were conducted on a very small scale. Otto Schott also produced some relatively striae-free optical glass at Jena at the end of the 19th century with the use of small, stirred, platinum-lined crucibles, but platinum was regarded as too expensive to justify its use in large-scale commercial optical glass melting. Instead, some remarkable improvements were made in the composition and construction of ceramic glassmelting pots. The most dramatic of these were made during World War I by Dr. A. V. Bleininger (one of the founders of The American Ceramic Society) when he was heading ceramics research at the National Bureau of Standards. Bleininger and his co-workers succeeded in developing a porcelain-like composition that was cast to make extremely durable glassmelting pots that greatly reduced striae.

After a great burst of activity during World War I to give the United States a domestic supply of optical glass for military use (binoculars, range-finders, periscopes, etc.) there was little done until World War II to exploit what had been learned. The National Bureau of Standards and Bausch and Lomb continued to produce small quantities of optical glass, but only after a national emergency again arose in 1939 did Corning

Glass Works, Pittsburgh Plate Glass, and others enter the optical glass business to fill government orders.

The potmelting process for producing optical glass remained in use at the beginning of the buildup to World War II, and this was a very inefficient procedure. After the molten glass had been melted and stirred, the pots were cooled very slowly and the glass allowed to crack into large chunks. The large pieces were broken down to smaller pieces and slumped in refractory molds, then pressed to roughly the shape desired before grinding and polishing. From 200 lb of glass came <200 lb of lenses.

In the early 1940s a group at Corning led by Dr. Charles DeVoe attacked the problem of mass-producing optical glass more efficiently. Platinum-clad melting and refining chambers as well as electric resistance heating were attempted in a series of small continuous melters, with each succeeding design incorporating improvements derived from the previous unit. At the same time, a long series of developments improved the performance of platinum-clad stirrers to attenuate the few striae that survived the melting and refining operations. Finally, in 1942, Corning achieved a workable system that delivered 50 lb/h of good-quality optical glass. Eventually it was found that platinum lining was required only in the refiner.

Fiberglass

Continuous Glass Fibers

by Doug Hofmann, Owens Corning

Even though glass fibers have been produced since ancient times, they remained the curiosity of dreamers and artisans until very recently. In the early 1930s Owens-Illinois Glass Company began to develop expertise in the area of glass-fiber production and in the late 1930s in collaboration with Corning Glass Works, developed a commercially viable process for the production of glass fibers, thereafter assumed by their descendant, Owens-Corning Fiberglas.

The first continuous glass-fiber production began in late 1937. Glass marbles of the newly developed "E" composition were manufactured by Akro Agate Company in Clarksburg, WV, and then remelted and fiberized into continuous fibers at the Owens-Illinois Glass Company Technical Laboratory in Newark, OH. As one should expect from any budding technology, the operation was minute in comparison with the processes as they currently exist, each position producing only about one pound of glass fiber per hour, dwarfed by the outputs achieved today of more than one hundred pounds per hour. Since those modest beginnings, the processes have been improved through changes in furnace design (marble remelt to direct melt, utilization of regenerative and recuperative firing, and electrical-melting technology), material improvements (refractory ceramics and metals), and advances in mechanical equipment and control technology.

Along the way, many different glass compositions have been attempted and tested, but by far the most widely utilized has been and continues to be E-glass. Originally developed to have good electrical resistance, its good tensile strength, high modulus of elasticity, and low density allow it to be used for other than electrical applications. E-glass is utilized as a general reinforcement for plastics of all types and as a general purpose fabric whose excellent strength, durability, and fire resistance make it attractive for everything from the covering for today's office dividers to the coverings for large football stadiums. Light-weight items resulting from a combination of the low densities of plastics and glass have had considerable impact on reducing fuel consumption in the transportation industry.

Continuous glass fibers are produced in many forms: yarns, which are then woven into fabrics; rovings, which can be wound onto mandrels for producing pipe or tanks, woven into coarse fabrics for making automotive and marine parts, pulled through dies along with plastic resins to produce ladder rails and the like, or used in gun spray-up applications where the glass is chopped and sprayed onto a form along with resin simultaneously to produce the hulls of boats.

The glass also can be produced in chopped form, either chopped from the dry product, as in chopped strand, or chopped as a part of the fiber-forming and packaging process (wet chop). These products can be used to reinforce plastics as well as to create glass mats that are the basis for modern shingle manufacturing.

Although E-glass dominates the continuous-glass-fiber market, other glass compositions have been developed through the years to meet various needs. Some of these glasses have proved their value, others have faded into obscurity, having been replaced by materials more suited for the particular application. C-glass has been utilized in battery separators because of its superior acid resistance in comparison to E-glass. ECRglas™, originally developed as a zero boron glass, has found its application in corrosion-resistant tanks and pipes, where resistance to water and acid and overall strength are critical parameters. S-glass, with its high tensile strength and modulus has found a home in aerospace, where high strength-to-weight ratios are critical. A special version of the S-type composition, Hollex™, takes advantage of hollow glass fibers to increase the strength-to-weight ratio even more. AR-glass has found some limited application in reinforced concrete. Even A-glass is now being formed into continuous fibers for reinforcement where water durability and electrical resistance are not critical.

Looking toward the future, a newcomer to the continuous-fiber market is Advantex™, which boasts tensile strength, modulus, and electrical properties approaching E-glass and corrosion properties exceeding C-glass and approaching ECRglas. It also practically eliminates fluorine and particulate emissions (at the source) that accompany the manufacture of most E-glasses, because the glass composition does not contain the volatile elements that contribute to such emissions.

Continuous glass fibers pervade our world: from the lowly fly-rod to high-voltage standoff rods; from automotive body panels to M-1 tank turrets; from electrical wire insulation to multilayer printed circuit boards for computers; from wall-board and shingles to the Silver Dome. Glass fibers have changed the basic building blocks of today's products.

Steam-Blown Glass Wool

by Charles Rapp, Owens Corning

Fibrous glass has been manufactured and used for thermal and acoustical insulation and for other purposes for more than 100 years. Most early fibers were produced by a "mineral wool" process in which rock or slag was melted in a cupola and fiberized by an air or steam blast process. Fiberizing was typically done by feeding a stream of glass approximately 0.325–0.5 in. in diameter to a steam nozzle at a rate of ~500 lb/h. Steam emitted from the V-shaped nozzle would intercept the glass stream at a 90 angle. The intense shear and turbulence produced by the V-shaped nozzle would accelerate and attenuate part of the molten stream into fiber. The wool produced typically contained as much as 50% "shot" and other unfiberized or poorly fiberized material. This large quantity of unfiberized material resulted in a poor production efficiency. Also, and even more importantly, the large-diameter "slivers" of glass that were produced would stick in the skin and make the product objectionable to handle. The wool was often granulated to remove the shot, which resulted in short fibers and poor product integrity.

In the 1930s a dramatic improvement was made in the quality of glass wool when an advanced process was developed by Games Slayter for producing steam-blown glass fibers. This new process removed most of the objectionable qualities of the old mineral wools. Various aspects of advanced glass technology were applied to the glass formulation, melting, delivery, and fiber-forming process so that a glass wool was produced with long fibers of well-defined diameters that was substantially free of shot or other objectionable nonfibrous materials. This led to the rapid acceptance of fiber glass as the preferred thermal and acoustical insulation for buildings and other applications. Rapid market acceptance and growth occurred during the 1940s, 1950s, and 1960s.

In the steam-blown glass-fiber process, relatively pure glass is melted from conventional batch materials, such as sand, soda ash, dolomite, limestone, and borax. After melting and fining, the glass is fed to an electrically heated platinum trough with drain holes ~1–2 mm in diameter. Glass flow rate per hole is quite low and is typically one to several pounds per hour per hole. However, holes can be quite closely spaced in rows so that total throughput can be high. Pairs of linear steam or air blowers are positioned directly below the holes, one on each side of the holes, and are pointed in a downward direction. Glass draining from the holes is captured by the blowers, accelerated downward, and is attenuated into long, thin fibers. Shot or unfiberized material is very low. This type of fiber produces a low-density mat with the high resilience now commonly associated with glass wool.

In the 1950s and 1960s the rotary process for producing glass wool was developed. The rotary process can increase throughput and is primarily used today to produce building insulation. However, the steam-blown process is so simple, low cost, and versatile that it continues to be used in many places to produce glass- and basalt-based mineral wool.

Rotary Fiberizing

by Neil Cameron, Owens Corning

Although the steam-blown wool process gained wide acceptance after its invention in the early 1930s and, in fact, continues in use in some parts of the world, there was continued interest in developing methods of manufacture that would produce even higher quality at greater throughput rates.

Centrifuging or rotary production methods had been available in Europe since the late 1920s when the Hager process was invented. This process, where molten glass dropped onto a horizontal rotating disk with fibers being flung off the periphery, produced fairly coarse fibers and was subsequently overshadowed by the steam-blown wool process. Centrifuging continued to be an area of interest for several companies, and, in the mid to late 1940s two companies, St. Gobain and Owens Corning Fiberglas, independently developed a method of rotary manufacture. This took a multitude of horizontal glass streams produced by flowing glass through a perforated rotating hollow cylinder (or spinner) and fed them into an annular, downward-firing flame that attenuated the glass streams into fine fibers. St. Gobain was awarded the patent on this method because of the earlier filing priority, and, in 1956, Owens Corning Fiberglas licensed the process from St. Gobain.

In the late 1950s Owens Corning developed and patented a different rotary method, where an annular burner was used to heat the spinner, and an annular steam blower was used to attenuate the glass streams into fibers.

Variants of these two basic rotary methods are the backbone of the glass-insulation industry today, and each year many hundreds of thousands of tons of product are made worldwide using both methods. From the early days when spinners measuring 8 in. across and containing a few hundred holes were used, the process has evolved such that current spinners up to 2 ft in diameter are now used, each containing more than 10,000 holes.

The global adoption of these high-throughput rotary fiberizing methods has resulted in the widespread availability of cost-effective insulation. This has revolutionized residential and commercial construction techniques and led to energy-efficient and comfortable buildings worldwide with an associated savings of energy equivalent to billions of barrels of oil as well as reduced pollution.

Glass Specialty Items

Glass-Ceramics

by Linda R. Pinckney, Corning Incorporated

Glass-ceramics are polycrystalline materials formed by the controlled crystallization of glass. Internally nucleated glass-ceramics were discovered in the mid-1950s at Corning Glass Works by S. D. Stookey, and they since have found numerous applications worldwide. Most commercial glass-ceramics are formed by highly automated glass-forming processes and converted to crystalline products by the proper heat treatment. Glass-ceramics also can be prepared via powder processing methods in which glass frits are sintered and crystallized.

Glass-ceramics provide a wide range of properties not readily found in conventional glasses or ceramics, such as thermal expansion coefficients that can range from -75×10^{-7}/°C to $+200 \times 10^{-7}$/°C. They possess highly uniform microstructures, with crystal sizes on the order of 10 μm or less; such homogeneity ensures highly reproducible physical properties. Glass-ceramic design allows for the properties of the parent glass to be tailored for ease of manufacture while simultaneously tailoring those of the glass-ceramic for a particular application.

Most commercial glass-ceramics are valued primarily for their superior thermal properties, particularly an ultralow thermal expansion over a broad temperature range coupled with high thermal stability and thermal shock resistance. Zero or near-zero expansion materials based on stuffed β-quartz crystals are used in high-precision optical applications, such as telescope mirror blanks and ring laser gyroscopes as well as for stove cooktops, cookware, wood stove windows, and fire doors. On the other hand, the combination of high mechanical strength and toughness and zero porosity is exploited by applications ranging from architectural materials and tableware to dental restorations and bone implants. Certain glass-ceramics containing fluorapatite can be bioactive: they have the ability to bond with bone. Other glass-ceramics based on fluormica crystals can be machined to high tolerances using standard metal-working tools. These versatile materials are well positioned for the next century, with ongoing research into potential uses for advanced flat-panel displays, solar panels, data storage devices, optical and optoelectronic devices, high-performance multilayer packaging materials, and refractory, corrosion-resistant coatings.

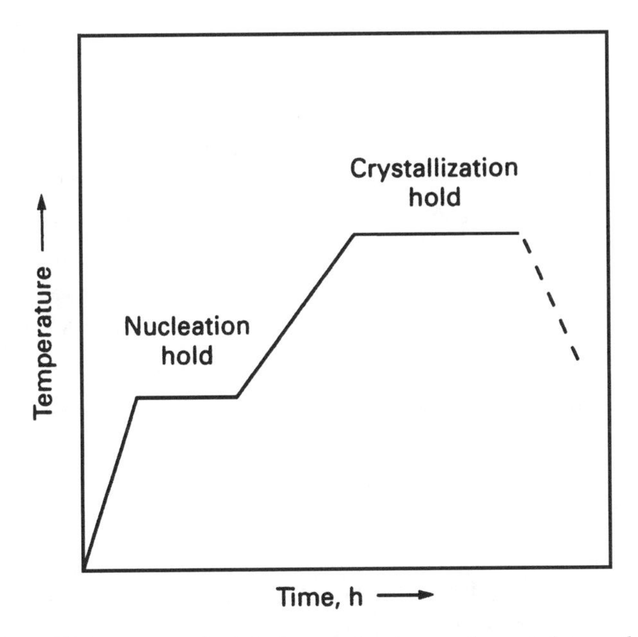

Figure 10. *Heating schedule for nucleation and crystal growth in glass-ceramics. Courtesy of Corning.*

Glass-Ceramics

by Aldo R. Boccaccini, Technical University of Ilmenau

Glass-ceramics are the result of the separation of a crystalline phase from the glassy parent phase in the form of small crystals resulting from special heat treatments to cause nucleation of very many fine crystals followed by growth until the crystalline phase reaches from ~50%–100% of the volume. The type, amount, size, and shape of the crystals are controlled and the properties of glass-ceramics can be varied over wide ranges.

Glass-ceramics were discovered by S. D. Stookey at Corning Glass Works in 1957 as a result of work on photonucleation of opal glasses. He found that special nucleating agents such as TiO_2 could initiate nucleation.

Areas of application include electromagnetic windows or radomes for guided missiles, where high hardness and strength, appropriate dielectric properties, and adequate thermal shock resistance were required. Low-expansion glass-ceramics are probably the best-known and most widely used, including compositions in the Li_2O–Al_2O_3–SiO_2 system. A variety of porcelain-like and transparent cooking and baking articles is manufactured from them.

Glass-ceramics are used in precision instruments, such as fluidic devices, magnetic disk recording systems, and many other applications in electronics, optics, and computer hardware where close dimensional control is required. The ability to shape complex parts by chemical machining makes possible intricate shapes, holes, etc. Machinable glass-ceramics constitute another important family. These are based on fluorine-containing mica crystals and can be machined to high tolerances without introducing catastrophic cracks. Glass-ceramics also are used in dental restorations. Architectural materials made from silicate wastes (e.g., incinerator flyash) constitute another important class of glass-ceramics, which offer an attractive solution to the problem of recycling this type of waste.

Radiant Glass-Ceramic Cooktops

by George H. Beall, Corning Incorporated

Radiant cooktops are currently the largest single market for glass-ceramics. The first glass-ceramics developed for smooth-top cooking were white materials based upon the low-expansion tetragonal lithium aluminosilicate β-spodumene. The nucleating agent was titanium oxide, which developed into the minor phase rutile during crystallization, and, with its high refractive index, accounted for the strong white opacity of the glass-ceramic. This product was originally marketed as "The Counter that Cooks." Unfortunately, resistive heating elements had to be separated from the underside of the cook-top in order to avoid electric short circuiting to the cookware caused by the high mobility of lithium ions at elevated temperature. This separation combined with the low thermal conductivity of the product produced mediocre performance.

An important advance occurred when a black infrared-transmitting product was developed. In this case, a nanocrystalline lithium-stuffed quartz glass-ceramic (hexagonal), somewhat similar to the transparent cookware VISIONS®, was developed. Nonscattering, tiny (~50 nm) β-quartz crystals were internally nucleated with great efficiency through the use of the nucleating crystal zirconium titanate ($ZrTiO_4$). Additionally, a coefficient of thermal expansion very close to zero was achieved with this material. In modern electric rangetops, a surface temperature up to 700°C can be produced under dry cooking conditions at a given burner position. Because the edge temperature is maintained at an ambient condition, hoop stresses would develop around the hot burner in a material with normal expansivity. This would result in undesirable edge tension. For maximum strength reliability, these stresses must be avoided, hence, the advantage of near-zero thermal expansion.

For both aesthetic and technical reasons, this β-quartz glass-ceramic was doped with vanadium to render a desirable black appearance in visible light. Nevertheless, the vanadium permits almost complete transmission of light in the deep red and near infrared. This allows radiant transmission from both standard heating elements and the novel tungsten–halogen elements now used for especially quick heat-up.

Radiant glass-ceramic cooktops are now widely accepted and compose more than two-thirds of the electric cooking surfaces presently sold. There are three major suppliers: Schott (Ceran®), Eurokera (Corning Incorporated/St. Gobain – Keraglas®), and Nippon Electric Glass (NEG – Neoceram®).

Glass Microspheres

by Warren Beck, 3M Company, retired

Glass microspheres, although largely invisible, are the key component in composite products that are having a major favorable impact on personal safety and the environment worldwide. Microspheres, ~1 μm to 1 mm in diameter, are large enough to be studied under a microscope but too small to be studied with the unaided eye. Their emergence as an important ceramic product is due to the unique properties that can be incorporated into them.

The first substantial use for glass microspheres began in the late 1960s when "beads" made from scrap window glass were coated on highway centerline paint to improve visibility. This "beads on paint" construction provided low efficiency retroreflection of light from automobile headlights at low cost and remains in use today. After World War II the growing automotive traffic created an urgent need for high-efficiency retroreflective traffic control signs. This need was best met by the development of plastic films incorporating, as the key component, a monolayer of millions of high-refractive-index optical lens elements (glass microspheres) per square foot. To function as a retroreflector it is required that

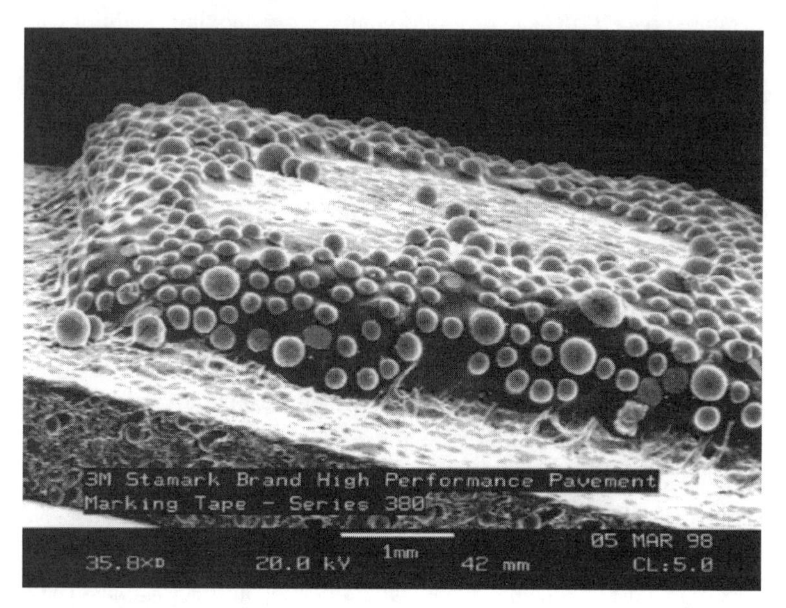

Figure 11. *Transparent microcrystalline ceramic beads used as reflective lens elements in reflective tape. Courtesy 3M Company.*

the sphere focus the light from oncoming automobile headlights onto a reflective concentric mirror surface at the rear of each microsphere so that a significant portion of this light is reflected and returned very close to the light source, making it highly visible to the driver of the automobile. To focus light in this manner spheres of 1.9–2.7 refractive index were required, depending on the refractive index of the medium in contact with the front and the spacing of the concentric mirror at the rear of the sphere. Durable, colorless glasses of these refractive indexes were unavailable prior to the mid 1940s. Fortunately, the small dimensions of the microspheres made possible a new technology area, that of rapidly cooled glasses, and this was pioneered by the 3M Co. This enlarged the possible composition range dramatically so that high contents of high-refractive-index network modifiers, such as titanium, barium, strontium, zinc, lead, bismuth, and other oxides could be incorporated. Only minor amounts of the traditional low-index glassformers, such as silicon, boron, and phosphorous oxides, are needed. The very low viscosities of the high-index melts permitted the development of new processing techniques. The microspheres are formed by several different processes, all of them having in common the steps of fusion and of rapid cooling from the melt. One exception is the process used to manufacture microspheres of superior hardness that can withstand the abrasive action of sand under automobile tires. Conventional glass fusion processes do not permit ready fusion of the refractory compositions of superior hardness; therefore sol–gel processing has been used to provide longer-lasting compositions for heavy-traffic areas. Some significant nonoptical uses have been developing for glass microspheres. In one such use with significant potential it is a major component of plastic composites to improve wear resistance and hardness with little loss of transparency.

Durable, low density (0.1–0.5 g/cm^3), high-strength hollow glass microspheres, also pioneered by 3M, became commercially available in the 1960s and have become an important industrial product. Their processing also depends on rapid cooling. It makes use of the well-known "reboil" phenomenon. Control of the process, gas content, and composition results in a fairly uniform wall thickness and density. The many uses of hollow glass microspheres are primarily in composites, where they are unseen but of great value because of their low density and high compressive strength. The major usage area is as a filler for plastics for aerospace, hydrospace, and highway transportation. The hull of the Alvin deep sea research vessel is filled with a buoyant composite of hollow glass microspheres in an epoxy resin polymer. In another significant area of use it acts as a sensitizer for slurry blasting agents to make them safer to use by making the explosion controllable. Cartridges incorporating the microspheres in a slurry of ammonium nitrate and reducing agent are replacing the former industry standard dynamite explosive.

Laminated Glass

by Paul S. Danielson, Corning Incorporated

The theoretical intrinsic strength of glass is almost never observed in practical applications, due primarily to the effects of flaws on the glass surface that provide break sources. Some technologies used to overcome the surface flaw sensitivity of glass include thermal tempering and ion-exchange strengthening. Both of these techniques have the effect of putting the surface of a glass article under compressive stress, thus neutralizing the effect of the presence of flaws. (Thus, in order for fracture to occur, the compressive surface stress must be overcome, in addition to the tensile stress required for a crack to propagate.) Another extremely effective, although narrowly practiced, strengthening approach is the use of a special glass–glass lamination process, such as the process used to make Corelle dinnerware.

In the glass–glass lamination technique, surface compression is generated by forming a layer of glass (or skin) having a lower thermal expansion coefficient over and around an interior layer (or core) having a higher expansion coefficient. At the forming temperatures, no stresses exist, but as the glass "sandwich" cools, the interior contracts more than the skin, creating substantial surface compression and protection against breakage.

In the Corelle dinnerware application, the skin glass is a transparent, highly durable glass, while the core is a densely opaque, less-durable glass. To solve the problem of coating one glass over another, the inventors of the Corelle process designed a mechanism to form a thin, laminated three-layer sheet of glass (skin–core–skin), which is vacuum formed into moving molds. At just the right temperature in the cooling step, the ware is trimmed out of the laminated glass sheet using a sort of "cookie cutter" process. The trimmed edge is fire polished, followed by decoration, and a final thermal treatment.

The resulting material is unusually strong, light, and durable. Corelle is the single most popular dinnerware in the world, with over a billion pieces having been made. However, the glass–glass lamination process has not been broadly applied to other product areas.

Borosilicate Laboratory and Consumer Glassware

by Paul S. Danielson, Corning Incorporated

Before the late 1800s the development of new glass compositions was an art, at best. However, beginning around the 1880s Otto Schott and others in Europe began the systematic study of the relationship between glass composition and properties, especially optical properties. One of the less-conventional elements to be incorporated successfully and beneficially into new glass compositions was boron, and, by the early 1900s several technical glasses had been developed with significant boron contents.

Two special attributes of boron-containing glasses that were exploited in the United States were relatively low thermal expansion coefficient and increased chemical durability. These properties were very successfully married in a composition used first for railroad signal lenses, acid battery jars, and laboratory ware. An even more commercially successful application, however, resulted from the discovery that the same composition served well for cooking vessels. Glassware sold under the Pyrex trademark and eventually others, such as Fire King and Kimax, became a staple of both laboratory and kitchen wizardry. Other similar compositions used the thermal and chemical durability of the borosilicates for lighting applications—lenses and reflectors in incandescent lamps and automotive headlights.

Large, Flat-Glass TV Tubes

by Yasuo Sato, Asahi Glass Company

Optical displays are essential to the information age. The glass TV tube (a form of cathode ray tube, CRT) is one of the important inventions of the past 100 years. Other new displays such as the LCD (liquid crystal display), PDP (plasma display panel), and FED (field emission display), have been developed, but the CRT continues to be widely used because of its resolution, brightness, and competitive cost.

Just as movies on large screens are popular, large-screen TV is desired by customers. Development of a larger TV with light weight was taken as a target to compete with large displays of the LCD, PDP, and FED types. The size of glass for TV displays has recently been increased from 14 to ~40 in. Tsunehiko Sugawara and Yusuke Ono developed a wide panel for the flat CRT, and its delivery began in 1996. Demand in Japan is booming and is expected to spread worldwide.

The flat glass bulb is delivered to CRT manufacturers, who have achieved many technical advances in electronics for the flat CRT. However, only the advances related to glass are described here. Because the glass bulb is an evacuated vessel, the flat panel suffers from a huge force, estimated to be ~6 tons from the pressure over its face. This causes tension on portions of the external face. The ideal shape to withstand this force would be a sphere. Achieving sufficient strength in a flat panel is very difficult compared with the ordinary, nonflat, curved panel. With previously existing technology, a thick, heavy, and less reliable flat panel is required.

In the new technology, physical tempering is applied in the process of pressforming the flat-glass panel. The principle of physical tempering has long been used to temper glass but seldom has been used for box-shaped glass. Newly achieved technology has made it possible to create an appropriate compressive stress in the appropriate area of the box-shaped panel to compensate the tensile stress caused by evacuation. The result is the first success in the world to produce a flat-glass CRT. This flat glass bulb with light weight, high heat resistance, high impact strength, and high reliability is expected to give a further lasting role to the glass TV tube.

Automotive Solar Control Electrically Heated Windshield

by Edward N. Boulos, Ford Motor Company

Ford Motor Company Glass Division developed the first electrically heated windshields using transparent conductive films used on production vehicles. There are no visible lines to distract vision as in the competing technologies of resistor wires embedded in the windshield or the resistor grid pattern used on electrically heated automotive rear windows.

The new product was first used in the windshields of 1986 Taurus, Crown Victoria, Grand Marquis, Continental, and Lincoln Town Car lines. A thin multilayer coating based on metallic silver and zinc oxide was deposited by planar magnetron cathodic sputtering on a glass surface that had previously been provided with fired, silk screened, silver enamel bus bars, and the coating was protected by lamination with poly(vinyl butyral) in the windshield. The bus bar designs were developed to optimize heating uniformity while minimizing hot spots. When installed in the vehicle and connected to the electrical system, this electrically heated windshield would clear 0.10 in. of ice or frost in 2–3 minutes at 0°F. The windshield still met the required ANSI federal safety glass standards, including 70% visible-light transmittance.

Transparent, multilayer electrically conductive films (transparent electrodes) were used in windshields and integrated into the vehicle electrical system for the first time in the automotive and glass industries. The silver multilayer film stack is an optically designed interference filter that maximizes visible-light transmittance and minimizes visible reflectance while achieving the required sheet resistance of 2–3 Ω/square. An additional benefit of this coated glass is solar load reduction by reflection of near-infrared radiation, which reduces the load on the vehicle climate control system. Compared to a conventional clear-glass windshield, the solar heat gain is reduced by 28%, and 40% of the sun's ultraviolet radiation is blocked. Pioneering research and development work was done in the area of poly(vinyl butyral) adhesion to coated glass and the effect of photocatalysis by ultraviolet radiation.

These automotive windows with transparent electroconductive optical coatings are a milestone in the industry and a forerunner of succeeding products, such as a solar control antenna windshield and of possible future products, such as automotive glazing in which the transmittance of light can be controlled electrically. The use of encapsulated, coated polyester films in laminated glass also was an industry innovation that has since led to a current solar control windshield product and to electrically switching "smart window" concepts.

Automotive Tempered Window 2.5 mm Thick

by Makoto Iwase, Asahi Glass Company

Environmental awareness has been growing drastically in the automotive industry in recent years. Weight reduction reduces fuel consumption and emissions. There is a great interest in reducing weight, even by a few ounces per part.

The area of glass used in automotive windows has continued to increase so that there is a focused effort to reduce the weight of the windows. Plastic glazing is a possibility for lighter windows, but because of the nature of plastics, it is quite difficult to use plastic windows under conditions that an automobile typically experiences. Glass has many superior characteristics, such as surface hardness, rigidity, chemical stability, and production cost. Chemical tempering is possible to give a thin tempered glass, but it is not applicable because of fragmentation regulations for automotive glass and high production costs.

The thickness of automotive body glass has been reduced as progress has been made in physical tempering technology for glass. However, the thinner the glass becomes, the more difficult it is to temper the glass physically. Until the 1960s physical tempering technology was limited to a minimum thickness of 5–6 mm. The technology progressed, and, in Japan, 4 mm tempered automotive glass was added as a category of JIS (Japanese Industrial Standard). In 1985, 3.1 mm thickness was added as a JIS.

Since then automotive glass industries have been competing hard with each other to develop the technology to temper glass thinner than 3.1 mm. It has been acknowledged that it is not so outstandingly difficult to temper 2.8 mm glass with some restriction of the glass area. By applying current equipment and technology to an area of 0.4–0.5 m^2, everybody could temper 2.8 mm thick glass.

Asahi Glass set out to temper 2.5 mm thick tempered glass for the rear window. It was then almost impossible to do this using existing glass-tempering technology and equipment. Asahi's team, led by Tsuyoshi Igarashi, General Manager of the Technology Division, went back to basic fluid thermodynamics, studied the principle of physical tempering, and found an idea that led them to invent unique quenching equipment.

By improving the design of the nozzles for the tempering air blast, Asahi reduced the minimum tempering thickness and made the process more efficient in terms of air pressure and exhaust. Asahi successfully started to produce 2.5 mm thick automotive regular-size rear windows in December 1997. This development contributes to improved automotive fuel consumption, and is potentially effective in saving energy in tempering glass in this industry in the future.

Photochromic and Photosensitive Glasses

by Roger J. Araujo, Corning Incorporated

In most cases, the absorption of a photon by an oxide glass causes no permanent change in the properties of a glass except for a small amount of heating. In some glasses, an excited electron can be trapped at a site other than its ground state and this causes a change in the optical properties of the glass. A glass exhibiting a permanent change in optical properties is called a photosensitive glass. One exhibiting only a temporary change is called photochromic.

A family of photosensitive glasses in which permanent photoinduced absorption has been observed is a set of glasses in which electrons are trapped by noble-metal ions, such as copper, silver, or gold. Heating is usually required for the formation of metallic colloidal particles in these glasses. Cerium is essential if exposure to ultraviolet light is desired to produce the photoreaction. It is not essential if more energetic irradiation, such as γ irradiation is utilized. In either case, the requirement for hole trapping is met only if the glass contains a sufficient density of nonbridging oxygen atoms. Colloidal metal particles can be used to nucleate the formation of crystalline silicate phases.

Photochromic glasses—in which there is photolysis of suspended minute crystals of silver halide or cuprous halide—are well-known. Glasses containing silver halides have been commercially available for ophthalmic applications for many years. The importance of cuprous ions to the hole-trapping process in these glasses is well understood. The fact that boron coordination varies with temperature as well as with the composition of the glass is important to the precipitation of the halides in both of these families of photochromic glasses.

Ceramic and Glass Foodware Safety

by Edwin Ruh, consultant, and Linda S. Geczi and Richard Lehman, Rutgers University

One of the most important contributions of ceramic technology during the past 100 years has been the development of vitreous technology that has ensured the safe use of ceramic and glass foodware. Two general approaches to foodware safety have followed nearly parallel paths during the past 100 years; the careful formulation, control, and processing of lead- and cadmium-containing glazes to assure low migration levels, and the formulation of a new category of glazes virtually free of the most toxic of the traditional constituents, principally lead.

The use of lead in ceramic foodware has an extensive history. According to written accounts and from the analysis of archeological artifacts, lead has been used in foodware from ancient Egyptian times to the present. The broad popularity of lead as a glass and glaze constituent results from the numerous processing and property characteristics it imparts to the vitreous state. The addition of lead oxide to silicate glass or glaze compositions lowers the fusion point, widens the processing range, reduces surface tension, and permits greater flexibility in formulating a composition to achieve the desired properties of low expansion, smooth surface, and high brilliance. Lead glasses and glazes are highly resistant to devitrification, have good chemical durability, and have the ability to heal body defects such as blisters, pinholes, drying cracks, and other defects of the clay surface.

However, if lead glasses or glazes are improperly formulated or fired, toxic amounts of heavy metals can be released via migration to food substances in contact with the defective vitreous surface. The problem of lead migration was recognized early in this century and an increasing level of research was focused on fritting of lead oxides to reduce the lead availability and on understanding the proper formulation of lead glazes to minimize migration by interdiffusion. This effort culminated in the 1974 Geneva Conference on Ceramic Foodware Safety at which state-of-the-art technology was presented and the groundwork was laid for broad international standards on the test methods and permissible limits of lead and cadmium release from foodware surfaces. At the close of the century, the ISO completed its third issuance of international foodware standards that are the foundation for regional standards and for safe worldwide production and trade of ceramic products among developed countries.

An alternate approach to minimizing lead migration is the total avoidance of lead in the glass or glaze composition. Initial studies on leadless glazes predate the century, but significant formulation efforts commenced during World War II when shortages of lead oxide prompted investigation of low-temperature opaque glazes. Substitutions were tested in which other fluxes replaced part or all of the lead oxide constituent. More recent emphasis on leadless glass and glaze development occurred in the 1980s and 1990s as environmental, occupational safety, and lead-migration regulations became stricter. Three approaches to leadless compositions have been pursued. Direct substitution of bismuth for lead is the most obvious and produces adequate results. However, bismuth can impart a yellowish color under certain circumstances, the supply of bismuth is limited, the price is high, and the toxicity of bismuth itself may be an issue. A second group of leadless compositions uses zinc and strontium to provide the necessary fluxing. These glazes are glossy and fire well, but color development is poor. A third approach is toward alkali borosilicate (ABS) formulations. These glazes rely on alkali borate fluxing, and a typical composition may contain ~10% B_2O_3 and 10% $(Li,Na,K)_2O$ by weight. The ABS glazes are becoming widely used, particularly on bone china, because of the high expansion of the ware, but significant problems remain with its use. Higher firing temperatures are required to produce a smooth glaze surface, the leadless glazes react less aggressively at the body interface, defect rates are higher, and decoration is difficult. Continued development will result in increased performance and acceptance of this system.

Overall, the two approaches to safe ceramic foodware outlined in this article have produced outstanding results and have made a major contribution to world health. These methods continue to be practiced and developed on a worldwide basis with the desirable result that toxicity issues related to ceramic and glass foodware are rapidly decreasing. Indeed the isolated issues of lead poisoning associated with faulty ceramic ware have virtually vanished, and, when such instances do occur, they result from a failure to comply with good manufacturing practice. During the next century, attention is likely to focus on the possible subacute toxicity of other glass and glaze constituents, as increased means of detection become available to enthusiastic regulators. One can anticipate obvious targets such as bismuth and barium, but also on more subtle effects associated with strontium, manganese, lithium, boron, and even zinc.

The authors wish to thank Robert Beals and William Spangenberg for their contributions.

4

Ceramics in the Processing of Other Materials

A major use of ceramics since the beginning of the age of metals has been as high-temperature thermal insulators and containers for molten metals. These uses were greatly stimulated and extended by the growth of the iron and steel industry in the 19th century and the development of the glass industry. Within the past century there have been quite a few major innovations that have greatly increased the length of furnace use between rebuildings. Typical refractories today include many specialty items that are higher in cost but are cost effective because they give much longer service than their counterparts of the previous century. Furnaces in general have been improved by the use of fiberous refractory insulation in both crystalline and glassy forms.

A major type of improved refractory for steelmaking is carbon-bonded magnesia for basic oxygen furnaces that has extended furnace campaigns to 20,000 or more heats. Another approach to extending furnace life is the use of refractory castables to repair hot furnaces. Another important advance in steelmaking is the refractory slide gate used to control pouring molten steel from ladles.

Glassmelting furnaces also have benefited from innovations in refractory technology. Fused cast refractories made of mullite, alumina, and zirconia in various combinations are used. Dense zircon shapes also are used, especially for specialty glasses. Electric melting of glass is greatly facilitated by refractory tin oxide electrodes.

Still another area of materials processing that has benefited from improved refractories is that of rotary cement kilns. Direct-bonded calcia-magnesia refractories with zirconia additions to improve thermal shock have been useful in steelmaking as well as cementmaking.

Ceramic molds for the casting of bronze statues and other objects have been used from antiquity. Ceramics for the casting of precision parts made from modern alloys have improved so much in the past century that they constitute an important innovation. A related innovation is the ceramic cores that are used for hollow castings.

Another and quite different use of ceramics in materials processing is that of abrasives and cutting tools. Synthetic materials, including aluminum oxide, silicon carbide, synthetic diamond, and cubic boron nitride, are widely used in the cutting and finishing of ceramics and hard materials in general. Important ceramic cutting tools developed in the past century include cobalt-bonded tungsten carbide and whisker-reinforced alumina.

Refractories

Basic Oxygen Process Refractories

by Richard Bradt, University of Alabama

Since their introduction nearly a half century ago, the ultrasophisticated magnesia/carbon composites that are used to line the metallurgical process vessels known as basic oxygen furnaces (BOFs) have undergone a series of evolutionary developments unparalleled in any composites. The result is an improvement in lining campaign life from <50 heats to almost 20,000 heats of steel. Instead of the linings lasting a week, they now last almost 10 years. To put these advances in perspective, one need only consider that the BOFs are operated between 1700° and 1800°C in a turbulent, reactive, thermal cycling environment. These refractories are dynamic composites at the highest level of performance, because they are continually changing from the molten metal hot face to the vessel's shell cold face during their utilization in the steelmaking vessels.

To be sure, the advances that have been made in these products are the combined efforts of refractories producers and the steelmakers, the latter modifying their practices by incorporation of advanced processing concepts such as slag splashing. Nevertheless, the advances made in the technology of the magnesia/carbon refractory composites alone merit review. To appreciate the change in those refractories it is perhaps significant to first examine the magnesia-carbon composite BOF refractory as it existed in the early days of the BOF and then the modern magnesia-carbon composite refractory.

The early magnesia-carbon composite BOF refractories were manufactured in what might be considered a rather obvious way. Utilizing the proper quality of magnesia with the correct lime/silica ratio, brick were pressed and fired, yielding a porous magnesia refractory. These brick were subsequently impregnated with molten tar/pitch in a vessel that was first evacuated then pressurized once the refractories were submerged into the molten hydrocarbon. As might be imagined, the resulting carbon levels from the pyrolysis of the hydrocarbon were only a couple percent of carbon. Multiple impregnations increased the carbon levels a little, but hardly justified the effort. These magnesia-carbon refractories did enable the production of steel, but are far removed from the dynamic sophisticated magnesia-carbon refractories which line today's BOFs.

The magnesia-carbon refractory in the modern BOF is a multiphase dynamic high-technology composite that is continually changing during its campaign life through the *in-situ* reactions of its numerous constituents. These many-phase refractories may contain several different magnesias (some sintered and some fused), different carbons (natural graphites and lampblacks), a high-carbon-yielding resin bond, and usually several different metal powders (aluminum, magnesium, and silicon and their alloys are common). Producers often also add their own versions of "magic" dust. It is not uncommon to have these constituents develop a half dozen or more phases in dynamic equilibrium during their campaign.

It is difficult to imagine the extreme complexity of the kinetics and thermodynamics of such a multiconstitutent refractory containing molten steel and operating above 1700°C. In fact it is such a complicated system that the roles of the various components remain controversial in many instances. The process technology of these advanced composite refractories has enabled a significant increase in the carbon content—levels of 30% are possible. The metal powder additives appear to enhance the hot face dense zone that many deem critical to the longevity of these brick in service.

The unparalleled technological advances of these magnesia-carbon refractories have enabled the BOF, the central process vessel of a modern integrated steel mill, to reach new heights of productivity and quality in the modern metal world. Every steel product has benefited significantly from the advances in these ceramics.

Resin-Bonded Magnesia–Graphite Refractories for Application in Basic Oxygen Furnaces

by John Ainsworth, North American Refractories Technical Center

During the 1960s integrated steel companies began to use the basic oxygen process for making steel. Over the decades, the process has become one of the dominant methods of steelmaking around the world. The basic oxygen process requires ~45 min to make a heat of steel as compared to the process previously used, the open hearth furnace, which took 3–4 h to make the same heat of steel. The basic oxygen process consists of charging a large vessel with molten metal from the blast furnace with some scrap metal and lime-based fluxes (slag). A lance is lowered to about 8 in. above the steel bath and oxygen is blown onto the bath. Impurities, such as phosphorous and sulfur, are removed and the carbon level is reduced to a specified level.

Magnesia-based refractories are used in the lining of the vessels. These materials have very good refractoriness and are among the most resistant to the molten slags formed in the basic oxygen steelmaking process. Initially, the magnesia refractories were either ceramically bonded by firing in a kiln or they were bonded with tars and pitches. The tar-bonded brick would form a carbon bond as the tar pyrolized during initial heating of the lining. The carbon bond did not oxidize because a coating would deposit on the surface of the lining.

In the mid-1970s a new concept in refractories for basic oxygen furnaces emerged, the magnesia–graphite refractory. This refractory contains magnesia grain and graphite, and it is bonded with resin. As with tar, the resin pyrolizes and forms a carbon bond during heat up of the lining. The graphite provides the slag resistance. It is not wetted by the slags, and it is believed that graphite, with the carbon, creates pressure within the brick to prevent slag intrusion. The graphite content of the brick ranges from 5% to 20%, depending on the properties that are desired.

Since the introduction of magnesia–graphite brick, there have been many improvements and refinements. The first significant improvement was the addition of metal powders or antioxidants, such as aluminum, magnesium, or silicon. It was found that small additions of these metals improved hot strength, slag resistance, and oxidation resistance. The effects of different metals on slag corrosion resistance are shown in the accompanying figure. A second major improvement was in the purity of the components of the brick. It was found that there were major increases in lining life when the level of purity of both the magnesia component and the graphite component was 99% or higher.

The crystal size of the magnesia grain is extremely important, because it affects the slag resistance of the grain. The smaller the crystal size, the lower is the slag resistance. In sintered grains, the crystal size is 60–80 µm. In fused grains, the crystal size may be 1000 µm or larger. Therefore, fused, high-purity magnesia grains are used when the very highest slag resistance is required.

The technology that has been developed for BOF refractories has found applications in other areas as well. Resin-bonded magnesia–graphite brick are used in electric steelmaking furnaces, and they are used in the slag liners of ladles that are used to transport molten steel within a shop.

The continuous improvement in both steelmaking parameters and refractories has resulted in the number of heats on a refractory lining increasing from ~100 in 1960 to 24,000 in 1997.

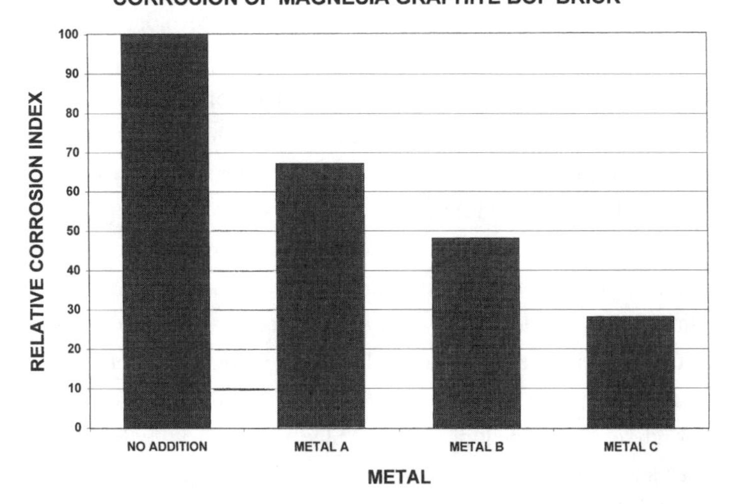

Figure 12. *Effect of metal addition on chemical corrosion of magnesia–graphite BOF brick. Courtesy of North American Refractories Co.*

Refractory Slide Gates

By Edwin Ruh, Rutgers University, retired

Nothing has revolutionized the steel industry as has the introduction of refractory sliding gates. Prior to their development, steel was poured by using stopper rods and nozzles. This method lacked precision and control required by new operating methods (e.g., continuous casting). The transition to slide gates gave better control over extended periods of time, proved to be more economical and dependable, and provided additional safety.

The sliding gate uses a mechanism that allows the control of steel flow to occur external to the vessel. Hence, refractory wear occurs only when pouring. The mechanism contains two refractory plates with predetermined bores. One plate is fixed and the other is stroked back and forth allowing for full close to full open operation, or any throttled position in between.

The concept of the sliding gate, developed by D. D. Lewis in the 1800s, did not come to the forefront until the early 1960s, when the quality of refractories increased to a point that made the sliding gate feasible. With the development of continuous casting, companies such as U.S. Steel saw the need for more reliable steel flow control. At their Monroeville research facility engineers took new refractory innovations and coupled them with the slide gate concept and developed a functioning system that blossomed into a business of its own in the United States under the name Flo-Com. At the same time competition came from Europe's Metacon and Interstop, Japan's NKK and Kurosake, and others.

Refractory manufacturers had to learn to develop a refractory that would not be wetted and that would be nonreactive to steel, and have a smooth texture so it would slide and not allow steel ingress into pores. The refractory also had to be manufactured to new, tighter tolerances required by the system. Refractory manufacturers also had to develop tests that would correlate to the steel plant operating conditions to be able to check quality and improve the performance of the refractory system.

A torch test was developed at U.S. Steel's Monroeville, PA, Research Center and provided a "go/no go" test simulating how it would behave in service. Initially, fire clay refractories were used, but these have given way to higher alumina contents, 85%–95%, and eventually magnesia- and tar-impregnated refractories. Today, plates are made in a variety of compositions that include zirconia and alumina carbons. Slide gates

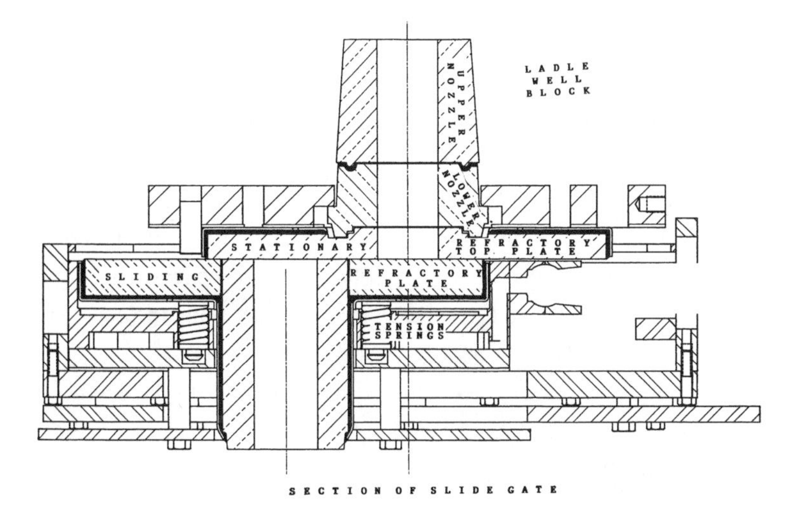

Figure 13. *Cross section of refractory slide gate. Courtesy of Flo-Con.*

have been applied to constant head tundishes, which now account for 85% of the steel cast. Slide gates have improved the making of steel and have been a contributing factor in keeping the price low. The change to slide gates made a new market with different demands for the refractories industry.

High-Purity Calcium Aluminate Cements

by Robert McConnell, Alcoa Industrial Chemicals

Portland cements are limited in the temperature they can withstand to ~650°C for constant temperature and to ~425°C for thermally cycled applications. This leads to the need for a bonding agent for monolithic refractories for molten-metal containment and for insulation to protect structural components in high-temperature processes. The binder must provide good strength prior to firing and through the entire temperature range from ambient to 1750°C. Hydraulic bonding, high-purity calcium aluminate cements (CAC) meet these needs.

The benefits of CACs were first investigated by Ebelman in 1848. The first commercial production of an intermediate purity CAC was in 1918, using bauxite and lime as raw materials. It was not until the 1950s that high-purity CACs were developed commercially. These high-purity

CACs allowed the development of monolithic refractories that are alternatives to pressed, fired shapes and to chemically bonded monolithics. See Table I for the typical composition of CACs. To achieve the low impurity level, the normal raw materials are a Bayer process calcined alumina and a high-purity limestone. The typical impurities found in high-purity CACs are SiO_2, Fe_2O_3, TiO_2, and MgO. The primary mineralogical phase in a high-purity CAC is CA with lessor amounts of CA_2, and $C_{12}A_7$. The primary impurity phases of concern are C_2S or C_2AS, C_2F or C_4AF, and CT. Phases are presented using the convention C = CaO, A = Al_2O_3, S = SiO_2, F = Fe_2O_3, and T = TiO_2.

The manufacturing process starts with cogrinding the limestone and alumina to a high percentage <45 μm (325 mesh), which intimately mixes the materials for a more consistent reaction process. The control of the calcia to alumina ratio is critical in this step. The ground components then are fed to a firing kiln to produce the CAC clinker. Although vertical-shaft kilns and melting furnaces have been used to produce CAC, most current manufacturing uses a rotary kiln to fire the limestone/alumina mixture to ~1425°C to produce sintered clinker. Complete firing is critical because uncombined CaO (free lime) above 0.25% can cause flash setting of the CAC when combined with water. The clinker is next ground into finished cement. Depending on the calcia-to-alumina ratio for the clinker and the target alumina content for the CAC, additional alumina may be coground with the clinker. The ground cement is typically packaged either in multiwall paper bags or in supersacks; both include a moisture barrier layer to minimize the deterioration of the cement that results from moisture pickup.

As mentioned previously the primary application for high-purity CAC is in monolithic refractories. The cement content varies based on the needs of the mix designer. It may be as high as 35 wt% or as low as 3 wt% or even as low as 1 wt% when combined with other bonding agents. See Table II for the ASTM classifications of refractory castables, which is based on the weight percent of calcia in the total refractory mix. Increased percentage of calcia normally results in increased cured strength, and increased alumina will increase the refractoriness of the fired monolith if all else is constant.

Because the initial (cured state) bonding mechanism provided by high-purity CACs is a hydraulic bond and not dry-out, care must be exercised in the first 24 h after casting not to allow the mix to dry out. To achieve optimal properties, high-alumina CACs should cure at 30°–35°C, but >20°C will produce good properties. After the cure is complete, the monolith is dried (usually at 110°C) to remove excess water prior to firing. Finally, the monolith is fired to or near operating temperature. The water of hydration is driven off between 350° and 650°C. Failure to completely dry out or to provide adequate time for moisture release during initial firing can result in explosive spalling due to trapped steam.

The use of high-purity CACs continues to grow each year as additional applications convert from brick to monolithic refractories. The major driving forces for this conversion are the improved properties of the monolithic refractories and, for most installations, their lower total installed cost.

Table I. Typical Chemical Composition for Calcium Aluminate Cements

	Composition (%)			
CAC Grade	Al_2O_3	CaO	SiO_2	Fe_2O_3
Low purity	39–50	35–42	4.5–9.0	7.0–16
Intermediate purity	55–66	26–36	3.5–6.0	1.0–3.0
High purity	70–90	9–28	0.0–0.3	0.0–0.4

Table II. ASTM Classifications of Alumina and Alumina–Silica Castable Refractories

Classification	CaO content (%)	Comments
Regular castable refractories	>2.5	>300 psi MOR for normal strength, >600 psi MOR for high strength
Low-cement castable refractories	1.0–2.5	
Ultra-low cement castable refractories	0.2–1.0	
No-cement castable refractories	<0.2	No significant CaO can come from the bonding agent

Advanced Refractory Castables

by Robert E. Fisher, consultant

Refractory castables using calcium aluminate (CA) cement as the binder have been part of the spectrum of products available to the refractory consumer since the 1930s. However, their generally poor physical properties as compared to brick precluded their use as materials of choice in most applications, even though castables are easier to install than brick. Then around 1970 low-cement (LC) castables first appeared. These products possessed substantially superior physical and mechanical properties to their conventional castable counterparts, and even outperformed refractory brick in many applications.

LC castables (and all the offshoots from this basic concept) depend on the use of deflocculants and other flow modifiers to intimately distribute the CA cement and other ultrafine ceramic powders within the matrix of the castable. The result is low water demand to achieve optimum placement consistency. This in turn means low porosity (leading to better resistance to infiltration and corrosion by slags and molten metals, etc.) and excellent mechanical properties (hot strength, lower brittleness, and better resistance to thermal shock). A corollary benefit of lower CA cement is less lime, improving an LC castable's resistance to certain aggressive slags.

At first, widespread use of LC castables was hindered by a lack of understanding about the mechanisms by which the additives modify flow, reduce the water demand, and control the hardening time. Castables are, by their very nature, installation-dependent materials of construction. Erratic behavior of LC castables during installation was the cause of many premature failures, causing both manufacturers and consumers to back off from using them in many situations where they would have been an excellent choice if they could be installed without problems. Because of this lack of know-how in 1980 only one company in the United States was successfully offering a diverse product line of LC castables.

Because of technical activities by refractories technologists in cooperation with the additive and ultrafine powder producers, steady progress was made in gaining a better understanding of the additive systems. The result of this cooperation was less-installation-sensitive, more-user-friendly LC castables. During the decade of the 1980s LC and ULC (ultralow cement) castables grew exponentially in use as they gained widespread acceptance—outperforming conventional castables and brick in a wide variety of applications.

Understanding the rheology of these fluid systems, including the effect of additives, has made it possible to control and modify the flow behavior and hardening time of the systems. During the 1990s this understanding resulted in several totally new classes of advanced castables. High-cement/reduced-water castables are being used that possess extremely high mechanical strength at temperatures between 600° and 1000°C. Self-flow castables are now available that do not need mechanical vibration for placement and that can be used to cast complex, intricate shapes. LC and ULC castables can be installed or conveyed using concrete pumps, allowing large-tonnage jobs to be placed in a short time. In the latest application of this pumping technology, a hardener/accelerator is added at the hose nozzle, allowing LC and ULC castables to be sprayed or shotcreted into place.

Fusion Cast Refractories

by Michael A. Nelson, Corhart Refractory Company

Around 1920 the glass industry operations, including those of Corning Glass Works, were transitioning from clay pots to continuous tanks lined with clay refractories similar to those used for the pots. Life of these furnaces was typically 6–8 months. Study of many crystalline refractory "stones" at that time showed them to be predominantly mullite crystals. It was reasoned that a refractory made from mullite would increase furnace life, allow higher melting temperatures, and reduce glass defects. Making mullite refractories would require extreme heat. Thus, Corning's Dr. Gordon S. Fulcher suggested electric melting using the raw materials as their own container.

The new fused mullite refractory was called "Adamant" and patents were issued to Dr. Fulcher. The first successful tank block was made in July 1922. Demand for these fused refractories soon increased although they were priced 5 times higher than clay blocks. Production continued in the Corning pilot plant until 1927 when it became apparent that a dedicated facility and company were needed to handle this growing business.

At this same time, Hartford-Empire Company also was working on refractories for the glass industry. In an effort to resolve patent issues as well as to meld Corning's technology and Hartford-Empire's access to glass customers, a joint venture was formed. Thus, Corhart Refractories (Cor from Corning and hart from Hartford) was created to manufacture "Electrocast" refractories. Corning owned 75% of the company.

Construction of Corhart Refractories in Louisville, KY, began in 1927 with the first blocks from this plant shipped in March 1928. In 1929 Corning Glass Works, in partnership with Cie de Saint-Gobain of France, formed L'Electro Refractaire. In 1930 Corning Glass Works licensed Asahi Glass Company of Japan to manufacture and sell fused cast refractories.

World War II emphasized the need for both increased production and higher-quality refractories. Through work started in late 1930s by Corhart Research and Development, the first zirconia-bearing fusion cast refractory product was introduced in 1942 under the name of "ZED." An improved fused alumina–zirconia–silica (AZS) product was patented and introduced in 1943. It was named "ZAC."

In 1953 L'Electro Refractaire created a Research and Development Center near its new plant in LePontet, France. In the following years this R&D center played a major roll in the evolution of fused refractories. The introduction of oxidized refractories is a typical example of this evolution (arc fusion patent in 1958). The oxidized products exhibit low glassy-phase exudation, low blistering, and result in less defects in glass. In the 1970s, L'Electro Refractaire changed its name to SEPR and proposed new innovations as fused cast Cruciform pieces for glass furnace regenerators. In 1987 Corhardt joined SEPR in the Industrial Ceramics Branch of Saint Gobain.

Today, Corhart/SEPR offers the glass industry the largest variety of fusion cast refractories, including fused AZS, with several levels of zirconia plus variations in other oxides; fused α-alumina; fused α–β-alumina; fused β-alumina; 94% fused zirconia; fused magnesia–chrome; fused chromia–AZS; and monoliths based on fused grains. Some fusion cast refractories also are offered to the metals industry and among them there remains a fused alumina–mullite refractory product that is a variant and descendant of the original Adamant compositon.

Fused Cast High-Zirconia Refractories

by Kisuke Tajima, Asahi Glass Company

Improvement of the quality of glass is an essential factor in enlarging markets, especially in electronic applications, such as in the liquid crystal display (LCD). Defects in glass are persistently requested to be smaller and fewer. Improved refractories can help solve the problems. The high corrosion resistance and low defect potential of fused high-ZrO_2 refractories have been accepted by the world's leading glass manufacturers.

Several factors have contributed to innovative high-performance fused high-ZrO_2 refractories (FHZ). Intensive study of chemical and mineralogical composition led to the manufacture of huge blocks of high-ZrO_2 content (>90%) by the fusion-casting process without cracking, despite the extraordinary contraction of ZrO_2 during cooling. This has led to new applications for refractories and contributed to the production of high-quality glasses needed by the electronic industries.

It is well-known that ZrO_2 and Cr_2O_3 refractories are highly resistant to corrosion by molten glass, but Cr_2O_3 has the disadvantage of coloring the glass once some corrosion takes place. ZrO_2 on the other hand has no disadvantage of causing coloring of the glass, but has a severe problem of cracking because of its extraordinary expansion caused by phase transformation. Y_2O_3 is well-known as a stabilizer of ZrO_2 but is easily attacked by molten glass and causes defects.

Dr. Masataro Okumiya at the Asahi Research Center began to study FHZ in 1980. He found an optimum ratio of Al_2O_3 to SiO_2 with the addition of a small amount of phosphate that enabled him to obtain high-ZrO_2 blocks as large as 300 mm × 600 mm × 1300 mm. In 1981 Asahi glass started sales of FHZ. Toshiba, Carborundum, and SEPR followed.

After the first-stage application to glass tanks, a new version of FHZ was developed. The new FHZ has electrical resistivity about 5 times higher than the first version. The higher resistivity is essential for electric melting of glass of high electrical resistance.

FHZ has been successfully used for a variety of glasses (non-alkaline, aluminosilicate, borosilicate, and alkaline–barium) for flat displays, such as LCD and plasma display panel (PDP), TV tubes, and photomasks. FHZ is expected to be more widely used for manufacturing special glasses in the future.

Fused Cast Refractories for Molten Glass Contact Application

by Stephen M. Winder, Monofrax, Incorporated

Modern glassmelting furnaces must operate under demanding conditions with respect to environmental impact and quality of the glass manufacturer's product. An optimum package of glass contact refractories must provide the greatest possible service life while presenting the lowest possible defect potential to the molten glass bath. These competing needs have led to the use of different glass contact refractory products in different parts of the glassmelting tank.

Fused cast AZS-based refractories are composed of alumina and zirconia crystals embedded in a glassy silicate matrix (AZS). These refractories have found wide application in molten glass contact, primarily because of the anomalous solubility behavior of zirconia and alumina in silicate melts. The presence of dissolved alumina in a molten silicate is found to dramatically decrease the solubility of zirconia. In glass contact service, diffusion of (typically) alkali and alkaline-earth elements from the glassy melt into the AZS refractory triggers dissolution of the structural alumina crystal phase. The resulting presence of an alumina-rich matrix phase around the zirconia crystals decreases their solubility. This results in the development of a kinetic barrier at the glass–refractory interface, composed of zirconia crystals in an aluminosilicate glassy matrix, which serves effectively in passivating the corrosion mechanism. Concurrent development of a physical boundary layer between this chemical passivation layer and the glass melt further reduces corrosion because of glass melt contact. Although this development of a zirconia-rich passivation layer is the root of excellent corrosion resistance in AZS, it also is a potential source of highly insoluble defects in the glass bath. AZS refractories are available in different grades with ~33%, ~37%, and ~40% zirconia for use in increasingly aggressive environments. They provide successful service in glass contact application as side-wall blocks, tank-bottom tiles, and high-wear applications, such as throat blocks.

Fused cast zirconia-based refractories are considerably more expensive than AZS refractories. They contain lower-glassy-phase content and usually provide considerably lower defect potential than AZS refractories, because the corrosion mechanism involves direct dissolution of the zirconia crystals without generation of a passivating layer. Fused cast zirconia refractory also generates much lower volumes of gas bubble than AZS, so that "seed" and "blister" defects are minimized. This refractory finds useful application in areas of the glass melting tank where physical boundary layers are thickest, such as in bottom tile, where it can provide higher corrosion resistance than AZS. Fused zirconia can be corroded more rapidly than fused AZS in areas of high glass flow, particularly in melts that are rich in alkali and alkaline-earth.

Fused cast alumina-based refractories contain essentially zero glass phase, provide low bubble generation, and can perform with lower melt contact corrosion rates than AZS when situated in the cooler areas of a glass melter. Glass contact application is found in refiners, channels, and lips, close to the finishing operation, where temperatures are lower and glass quality demands are higher. At the higher temperatures usually experienced in the glassmelting tank, these refractories provide less glass contact corrosion resistance than AZS or zirconia.

Fused cast chrome-based refractories can offer extreme corrosion resistance, because of the very low solubility of chrome in molten silicates. These materials find use in contact with high-temperature, highly corrosive melts, but are limited from broad application because of their potential to cause coloration of the glass product.

Fused Cast Refractories for Glassmelting Superstructure Application

by Stephen M. Winder, Monofrax, Incorporated

Modern glassmelting furnace superstructure refractories must provide the greatest possible service life while presenting the lowest possible defect potential to the molten glass bath. The service environment may present a range of corrosive agents to exposed refractory surfaces in the form of vapors (NaOH, KOH, PbO, HBO_2, etc.), batch dust particles, and liquid phases in the form of condensates or run-down from overlying structure. Any interactions between the refractories and the corrosive agents may cause refractory corrosion and have potential to generate defects in the glass product.

Fused cast alumina-based refractories contain essentially zero glass phase and are widely available in α–β-alumina or sodium β-alumina ($Na_2O{\cdot}9Al_2O_3$) types. Both alumina refractory classes generate a thermodynamically stable surface layer of α-corundum crystals during superstructure service in the melting of most glass compositions. The α–β-alumina refractory has previously been successfully applied in furnace crown application at temperatures exceeding 1800°C and is now enjoying wide application in oxy-fuel furnace crown service. In the absence of contact with batch dusts, the fused cast alumina refractories offer supreme corrosion resistance and present extremely low (if any) defect potential to the molten glass bath because of the absence of liquid formation. Application of the β-alumina refractory is preferred in any areas of the superstructure where significant condensation of alkali species is expected (e.g., ports) or in areas subjected to severe thermal gradients (e.g., tuckstones). In situations of moderate batch dust contamination αβ-alumina alumina–zirconia–silica provide superior corrosion resistance to alumina–zirconia–silica (AZS), but consistently heavy batch dust contamination adds silicate liquid to the surface of the refractories, allowing dissolution of alkali, etc., from the furnace atmosphere. This presence of alkali-rich silicate liquid may cause refractory degradation by dissolution of the alumina and represents some potential for generation of aluminosilicate defects in the glass bath.

Fused cast AZS-based refractories are composed of alumina and zirconia crystals embedded in a glassy silicate matrix. These refractories have found wide application in molten glass contact, primarily because of the anomalous solubility behavior of zirconia and alumina in silicate melts. The presence of dissolved alumina in a molten silicate is found to dramatically decrease the solubility of zirconia. However, in glassmelting superstructure service, diffusion of alkali from the glassmelting atmosphere triggers dissolution of the structural alumina crystalline phase in the AZS. In the melting of complex TV glasses or borosilicates, the presence of other cations accelerates this corrosion phenomenon. The corrosion reactions result in the formation of a large volume of aluminosilicate liquid containing zirconia crystals at the AZS refractory surface. This liquid corrosion product is generated throughout the furnace campaign and is a potent source of defects, because of drippage and run-down into the glass bath. It is often mistakenly identified as exudation, which is actually a different mechanism for AZS refractory to form defects in the absence of corrosive cationic species. AZS commonly finds superstructure application in areas where heavy batch dust contamination is considered problematic for the fused cast alumina refractories.

Figure 14. *Fused cast alumina for oxy-fuel crown application. Courtesy of Monofrax, Inc.*

Dense Zirconia Refractories

by Allen D. Davis Jr., Corhart Refractories

The invention and commercialization of Pyrex® and other borosilicate glasses opened a whole new field of glass compositions with unique and exciting properties. Glasses could be used at higher temperatures and survive sudden chilling without shattering. Borosilicate glasses also could be more resistant to chemicals, stronger, more colorful, etc. Applications for glass that were previously unimaginable became possible. Simply put, glass became a technical, engineered material.

The problem, however, was that the traditional clay refractories in which these new glasses were melted were being attacked and destroyed even more rapidly than usual. It also became more difficult to make products that did not contain defects caused by the corrosion of the refractories. This led glassmakers to look beyond the traditional refractories and to explore new refractory compositions and new ways of manufacturing them. One discovery of this research was that zircon was remarkably resistant to certain of these new glass compositions. It also was known that the more dense and the less porous a refractory was, the more resistant it was to corrosion by glass. Conventional pressing and firing of zircon brick, however, could not provide sufficient density to maximize this advantage.

In the 1930s Corning/Corhart scientists developed a method for slip-casting the relatively large zircon shapes required for commercial glass-melting. Slip-casting is a process where solids, suspended in a liquid, are cast into plaster molds. The porous plaster draws away most of the liquid and leaves formed green shapes ready for drying and firing. This process was old and well-established for clay-containing materials; it was not for nonplastic materials, such as zircon. The Corning/Corhart process was notable because the casting slurry was water, not acid-based, and contained up to 80% solids. Using slight compositional adjustments to improve sintering, the resulting fired zircon ceramic was extremely dense and impermeable.

The greatly extended furnace life, which dense zircon blocks permitted, contributed considerably to making many of these exotic glasses both practical and affordable. Examples of glasses that have been manufactured using dense zircon linings include borosilicate tableware, pharmaceutical and laboratory ware, light bulbs, textile and reinforcement fiberglass (E-glass), and aluminosilicate glasses, including glass-ceramics for bakeware, ceramic cooktops, missile nose cones, and aircraft radomes. Today, textile and reinforcing fiberglass represents the largest market for dense zircon refractories. E-glass and related glass fibers are used in a myriad of applications, including the structural reinforcement/filler in molded or extruded plastics of all types; the fiberglass fabric in marine, automotive, and aircraft bodies and other structural components; tire cord; printed circuit boards; and other electronic component reinforcement, roofing fabrics and shingles, geotextiles for lining ponds and landfills, filters for air or liquids, and flame-resistant fabrics for home and industry.

In the intervening years, isostatic pressing, pioneered by Corhart, has allowed larger blocks to be manufactured while retaining or improving the density of the materials. Iso-pressing is accomplished by placing the powder or a preformed shape into a sealed rubber bag that is immersed in water or some other fluid. When the fluid is pressurized, the piece is uniformly compacted by the uniformly distributed pressure. The resulting compact is fired and diamond machined, similar to the slip-cast versions.

Iso-pressing also has allowed the development of new, dense, coarse-grained zircon refractories that have outstanding resistance to thermal shock in comparison to the poor thermal shock-resistance of the original fine-grained versions. Today, iso-pressing is the predominant manufacturing process around the world. It is not unusual to find several dense zircon products being used in different locations within a furnace.

Chromic Oxide Refractories

by Allen D. Davis Jr., Corhart Refractories

During the 1930s primarily with the leadership of Corning Glass Works, there was considerable research directed at developing improved refractories for glassmelting. One way of identifying candidate compositions, those materials not easily dissolved by molten glass, was to examine the crystalline defects called "stones" that occur in glass. One of the most persistent stones in a broad range of glass compositions is chromic oxide (Cr_2O_3), which also is called "chromia." Using synthetic chromic oxide, Corning/Corhart scientists began development of a chromic oxide refractory.

One way of forming relatively large, dense, compacted ceramics at that time was slip casting. In slip casting, the ceramic raw materials are suspended in fluid, usually water, then poured into a plaster mold with the desired shape. The fluid is drawn into the pores of the plaster, leaving behind a damp, but compact shape that then can be dried and fired. As a nonplastic material—that is, one that contained no clays—chromic oxide was a particularly difficult material to slip cast. Corning/Corhart scientists finally developed a nonaqueous solvent system that worked.

Firing the chromic oxide so that it sintered well also proved to be difficult. Sintering is the process by which the individual ceramic particles coalesce, bond together, and shrink together to densify into a strong, single piece. After many initial failures, persistent researchers learned the secrets needed and developed a strictly controlled firing schedule that remains proprietary to this day.

Around the same time, glass chemists developed a strong and durable glass composition ideally suitable for making glass fibers. Known as "E-glass," this composition was the basis for the worldwide textile and reinforcement fiberglass market that exists today. Chromic oxide was found to be an excellent refractory for containing this glass. Its corrosion was minimal, it caused few defects that could break the fiber during the drawing process, and its tendency to give the glass a slight green tint was unimportant in this application. The commercialization of chromic oxide refractories was critical to permitting continuous fiberglass melting furnaces to replace the older, two-step melting process of first making marbles, then remelting them to draw fibers. E-glass is used as a filler and reinforcement in all types of molded resins and plastics. Examples include the fiberglass fabric in marine, automotive, and aircraft bodies/structural components and printed circuit board reinforcement. Fiberglass also is used in tire-reinforcing cord, roofing fabrics and shingles, geotextiles for lining ponds and landfills, filters for air or liquids, and flame-resistant fabrics for home and industry.

Over the past few decades, Corhart introduced isostatic pressing, which has surpassed slip casting as the predominant means of forming chromic oxide blocks. Isostatic pressing involves sealing the ceramic powder in a flexible container that is immersed in a fluid. When this fluid is pressurized, the powder is compacted more uniformly than it ever could be with a conventional pressing operation. Besides permitting even larger blocks to be made, this process has allowed the development of thermal shock-resistant variants of dense chromic oxide. Modern fiberglass furnaces typically use two or more types of chromic oxide refractory products in specified locations throughout the melting and forming sections of the furnace. In the past 10 years, chromic oxide refractory use has expanded beyond its use in E-fiberglass to applications in extremely corrosive insulating fiberglass compositions and in furnaces for encapsulating and vitrifying industrial, municipal, and nuclear wastes. Chromic oxide refractories also are currently being used in the highest-wear areas of conventional glass container and flat glass melters as part of an industry-wide program to increase furnace productivity and to extend their useful lifetimes.

Doloma Zirconia Refractories

by Don Griffin, Baker Refractories

Direct-bonded doloma and magnesia doloma are used worldwide as refractory linings in rotary cement and lime kilns, steel ladles, vacuum oxygen decarburization units, and argon–oxygen decarburization (AOD) vessels. Fired doloma brick are produced using high-purity doloma grain. The doloma is comprised of 58% CaO, 40% MgO, and 2% impurities (Al_2O_3, SiO_2, and Fe_2O_3). The doloma grains are comprised of rounded magnesia crystallites embedded in a continuous matrix of lime. It is the CaO phase that provides the direct bonds after firing and is responsible for the chemical properties of the brick. The unique properties (i.e., high refractoriness and chemical compatibility to basic environments) of doloma brick make it suitable for these applications.

One of the major drawbacks in the use of direct-bonded doloma brick was its susceptibility to damage by thermal shock. A major difficulty in improving the thermal shock properties of fired doloma brick was identifying additives that were compatible in a lime-rich system but would not degrade other properties (hot modulus of rupture, slag resistance, volume stability, etc.) of the refractory. In magnesia refractories, chrome and spinel have been used to improve the thermal shock properties. These materials could not been used in fired doloma brick because of reaction with the lime in the doloma, which had a negative effect on the chemical and physical properties of the refractory.

Extensive research was conducted in the 1970s regarding the use of zirconia as an addition to other oxide systems to enhance the thermal shock resistance. Based on the successful use of zirconia in other oxide systems and its chemical compatibility with doloma, a development program was started in 1983 at Baker Refractories to determine the effect zirconia additions would have on the thermal shock properties of fired doloma brick

The addition of zirconia to a doloma brick results in improved thermal shock resistance through the formation of microcracks. During firing, microcracks form as a result of reaction between the lime in the doloma and the zirconia-forming calcium zirconate. A microstructure of a doloma–zirconia brick is shown on the adjacent figure. This reaction is expansive as evidenced by the increase in porosity of the zirconia grain and produces the microcracks in the brick. Theoretically the zirconia (specific gravity of 5.73) to $CaZrO_3$ (specific gravity of 4.78) reaction would result in a volume expansion of ~16.5%. A secondary mechanism of microcrack formation is the mismatch in thermal expansion between $CaZrO_3$ and doloma.

In 1984 the first trials of fired doloma zirconia brick were conducted in the burning zone of a rotary cement kiln. The trials were successful and later led to the application of this product in steel ladles and AOD

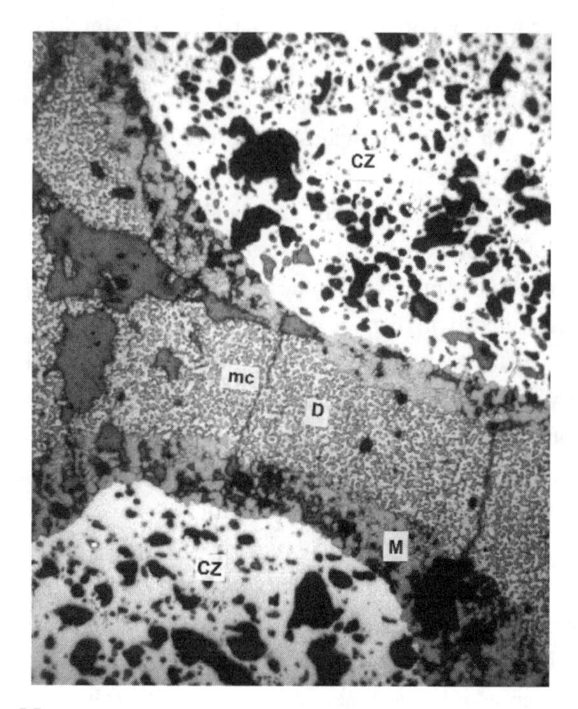

Figure 15. *Microstructure of a doloma zirconia refractory (CZ is $CaZrO_3$, D is doloma, M is magnesia, and mc is microcrack. Courtesy of Baker Refractories.*

vessels. The use of zirconia in doloma-based refractories has enabled the refractory manufacturer to produce products with low porosities, high hot strength, and high densities, while maintaining good thermal shock resistance. This combination of properties has expanded the use of doloma into high-wear areas in rotary cement kilns, steel ladles, and AOD vessels, thereby allowing the user to optimize refractory lining performance.

Electric Melting of Glass with Molybdenum or Tin Oxide Electrodes

by Ron W. Palmquist, Corning Incorporated

The first successful all-electric furnace was built by Cornelius in 1925 for the production of green and amber bottles. It used massive pure-iron electrodes, which corroded rapidly despite external cooling. Furnaces using graphite electrodes were built for the production of soda–lime glass, and proved most useful for the melting of colored glasses.

However, the increased use of molybdenum electrodes in the 1950s helped promote a rapid growth in the number of furnaces using electric power. Molybdenum electrodes do not produce fine bubbles or discolor the glass as does graphite. Although many gas- or oil-fired furnaces were boosted by adding some electric power, the design of efficient all-electric furnaces by Penberthy and Gell coincided with a relative decrease in the cost of electricity. Molybdenum electrodes allowed electric melting to be extended to higher-quality glasses, such as borosilicates for laboratory ware and, high-temperature aluminosilicates for fiberglass.

Although electric power remained more expensive than fossil fuels, an electric furnace was cheaper to construct because of the lack of regenerators. What would prove to be a special advantage in the late 1960s, an electric furnace with a cold top produced essentially no pollution. The cold top completely covered the molten glass with cold batch materials, and any volatile substance would condense in the batch before it could escape. Furthermore, limitations on allowed emissions fostered the greater use of boosting in gas-fired furnaces in order to increase output without increasing the emissions from a furnace.

The introduction of tin oxide electrodes in the 1960s further extended the use of electric power to glasses containing lead or other easily reducible oxides. A tin oxide electrode did not reduce the lead nor discolor the glass. The high volatility of lead was being recognized as a severe pollution problem, and a cold top electric furnace could eliminate this source of lead pollution. The reduction in volatility greatly benefited the crystal industry, and allowed the production of optical and specialty glasses with better composition control.

Although the energy cost for electrically melted glass in the United States always has been higher than that of a gas-fired furnace, by the 1980s some amount of electric boosting had been added to many gas-fired furnaces. Electric power is now considered as a means to control convection and temperature in a furnace and, hence, boost output while achieving better control of quality. Even float furnaces producing high-quality soda–lime glass now tend to use some electric boosting to enhance performance. With the technology now relatively mature, the relative amounts of electric and gas energy to be used in a furnace can readily be decided based on economic factors and pollution requirements.

Refractory Tin Oxide Electrodes

by Allen D. Davis Jr., Corhart Refractories

The use of lead in glass compositions causes these glasses to more effectively bend light rays. The results include the familiar sparkle of lead crystal and more powerful lenses in eyeglasses or precision optics. Lead also provides an effective shield against X-ray and nuclear radiation in everything from color TV tubes to windows in radiological laboratories and nuclear facilities. The addition of lead often optimizes the melting and expansion properties of glasses used for sealing light bulbs or electrical components. For centuries, lead-containing glasses have been prized for their beauty or for their practical properties. In many cases, nothing else does the things that lead does.

Lead has two problems. First, it is a relatively expensive material. The lead oxides in the glass batches can be easily reduced to lead metal in the flames of a furnace. This liquid lead sinks to the bottom and is wasted. By a process called "downward drilling," these metallic beads also enhance the corrosion by the glass of the refractories lining the bottom of the melter and shorten the life of the furnace. Furthermore, lead dusts or lead vapors can be lost up the smokestack. Unfortunately, this also relates to the second problem: lead is a toxin that can adversely affect human health.

The development of electric glassmelting was an exciting one for lead glass makers, because it promised to substantially reduce both the reduction of lead to metal and the suspension of dusts and vapors in the furnace exhaust. Trials with the molybdenum electrodes that worked so well in other glasses, however, showed that these electrodes reduced some of the lead to metal and also caused ugly, dark streaks in lead glasses. Researchers had found that tin oxide (SnO_2) was resistant to corrosion by molten glass and, with proper additives to enhance its electrical conductivity, was suitable as a semiconducting electrode that did not discolor lead glasses. Corning was an early user who developed many of the required electrical systems and connections, and, through its relationship with Corhart, also developed one of the first processes for manufacturing dense tin oxide refractory electrodes.

Today, tin oxide electrodes are, almost exclusively, manufactured by isostatically pressing the appropriate powders to the approximate shape, followed by firing, and diamond machining to the final dimensions. Because tin oxide is a refractory, it is normally embedded as an integral part of the normal refractory structure. It can be flush with the wall or protrude into the glass. Because it is a semiconducting ceramic, tin oxide also tends to operate at lower electrical current levels than conventional metallic electrodes. It can operate with or without the water-cooling jackets that are so critical for metal electrodes. These features make it an ideal electrode to help regulate and homogenize the temperature in the glass just prior to the forming of the glass objects. Tin oxide electrodes, therefore, are being used in the "conditioning" and forming sections of furnaces melting a variety of conventional soda–lime, borosilicate, and other specialty glasses in addition to its original use as the primary melting electrodes in lead-containing glasses.

Refractory Insulating Fiber

by Jayne M. Webb, Unifrax Corporation

In 1942 a Carborundum Company research scientist, John C. McMullen, invented a new refractory fiber when he melted aluminum oxide and silica in an electric furnace and subjected the molten stream to an air jet. The result was a cottony mass of fibers that could withstand temperatures of up to 3000°F. Almost immediately, it was found that ceramic fiber, while useful in its bulk form, was capable of being fabricated into a variety of beneficial products. These lightweight fiber-based products provided high-temperature resistance, low thermal conductivity and heat storage, immunity to thermal shock, good chemical and erosion resistance, and relative ease of installation. During the 1950s refractory ceramic fiber (RCF) was proved to be successful in a range of applications, and a limited line of Fiberfrax7 products was introduced to the emerging industrial market.

By the 1960s demands for a variety of high-temperature insulating, gasketing, and packing materials led to the development of a more complete line of ceramic-fiber products including papers, felts, blankets, board, ropes and braids, cements, vacuum cast shapes, and woven textiles. With the expanded product line offering, RCF layered furnace-lining systems were now being used in ceramic plants, steel mills, oil refineries, chemical plants, non-ferrous operations, and a variety of secondary heat-treating facilities.

The energy crisis of the 1970s and the corresponding increase in fuel costs opened the door for energy-saving RCF products in virtually all industrial segments. The insulating advantages were proved and the overall acceptance of ceramic-fiber linings was growing. The opportunities for ceramic fiber seemed infinite, and Fiberfrax7 products were being produced in 10 separate manufacturing facilities on six continents.

By 1980 after experimenting with various fiber chemistries, the temperature capabilities of ceramic-fiber board, blanket, and module products were expanded to as high as 2750°F. The increased temperature capabilities were achieved by blending a high-alumina or mullite fiber with standard alumina–silica fibers. The resulting materials were used on their own or as the hot-face in a layered insulating system.

Health concerns related to fibers in the 1980s sparked RCF manufacturers jointly to develop and institute a comprehensive product stewardship program to control employee and consumer exposures to airborne fibers. As industry leaders, they committed to reducing the potential for airborne fibers, controlling the size of fibers (reducing respirability), and enhancing the dissolution rate of fibers in body fluids. This commitment will direct development of new fiber chemistries and product forms that will continue to provide innovative solutions for high- temperature problems well into the next century.

In the 1990s this versatile ceramic product form was finding new markets and spreading to new parts of the world. Ceramic fibers found application in automotive catalytic converters, airbags, and fire protection in commercial buildings. Expansions into Southeast Asia, India, and China made an already global business bigger.

In summary, the invention of RCF has had a significant impact on reducing energy cost and improving productivity in a wide range of applications. Today RCF is crucial to many industrial applications as well as automotive, aerospace, fire protection, and appliance applications that affect our daily lives.

Metal Processing

Ceramic Shell Mold Investment-Casting Process

by Mehrdad Yasrebi, David H. Sturgis, and Karl Taft, Precision Castparts Corporation

The process of casting molten metal into ceramic molds has been used for centuries. Despite the popularity and wide use of metal casting, ceramic molds had neither the required refractoriness nor the precision for high-quality part manufacturing. These poor characteristics traditionally limited metal casting to applications where accurate dimensions or metallurgical properties were not required. Precise, high-quality metal parts had to be obtained by forging or simply by machining a part from a wrought block. Yet such processes are expensive, labor intensive, and in many cases impractical.

In the mid-1940s the development of the gas turbine engine for new aircraft generated renewed interest in the investment-casting process. It was found that many of the cobalt alloys used in the dental prosthetics industry performed well as turbine blades and vanes in these engines, and the investment-casting process, also used in dental casting, could be modified to produce these components. Parts could be made to close dimensional tolerances without machining, in high volume, and at comparatively low cost. This technology made a significant contribution to promoting the gas turbine engine as a viable, economical system for both military and commercial applications. Almost every blade and vane as well as many structural parts in aircraft engines are manufactured by an investment casting process. In addition, investment casting has a sizable market in the production of automobile parts, computer and electronic components, prosthetic appliances, valves and fittings, armaments, hand tools, and leisure equipment.

Investment-casting ceramic shell molds improved many of the existing shortcomings of commonly used metal-casting molds. The first step in the investment-casting process is to injection mold a precise replica of the part from a blend of polymers. Once the polymer replica in made, it is dipped into a ceramic slurry. The slurry contains a refractory system and, in most cases, a condensable binder, such as nanometer-sized colloidal silica or partially hydrolyzed ethyl silicate. The slurry coating is then sprinkled with a coarse refractory powder and allowed to dry. This alternating slurry and refractory process is repeated many times to produce a graded mold or investment. The above procedure allows for the fabrication of a mold that is a precise mirror image of the injection molded polymer. Furthermore, flexibility exists in changing the composition of each layer, allowing for the fabrication of a highly refractory mold with excellent surface finish, which is still cost competitive. Finished molds are then normally autoclaved to melt out the polymeric material, leaving a hollow shell. After metal casting, the ceramic shell is removed through mechanical or chemical methods to obtain the final part.

Investment-casting shell molds have stringent property requirements. Often, investment-casting slurries may remain in use for weeks or several months and should sustain no loss of properties. Molds require substantial green strength, which should be retained as the shell is heated up in the autoclave chamber. Shells should neither densify nor creep throughout the process. These requirements are challenging, because molds may experience large tensile stress during manufacturing as well as in the metal-casting process. In addition, molten metal may remain at temperatures in excess of 1500°C for many hours.

It is estimated that the present worldwide sales value of investment castings exceeds $5 billion and is expected to continue growing at a rate of more than 6% annually.

Ceramic Cores for Investment Casting

by Stuart Uram, consultant

The development of the jet engine during World War II required the manufacture of turbine blades for the hot section of the engine. These blades were produced from cobalt- and nickel-based superalloys that are very difficult to form by traditional metal-working techniques. The preferred method turned out to be the lost-wax investment-casting process. In this process, a wax pattern or replica of the desired part is formed in the first stage. The wax part is then "invested" in a ceramic material and the resultant ceramic mold is fired to a high temperature to burn away the wax and cure the mold. Liquid metal is poured into the mold and occupies the space, that was once wax. The net result is a metal part the exact shape of the wax pattern.

It became highly desirable to form hollow metal turbine blades. Initially, the hollow blades were needed for weight reduction and later cooling schemes were invented so that these turbine blades could be operated at very high temperatures for improved efficiency. Producing a hollow part was a technical challenge because it means that ceramic material must be able to form the inside of the part in the mold-building process. Because the passageway inside these blades is very small, of the order of 0.5 mm in many cases, the mold-building step could not be conducted in a reliable fashion. It became obvious that if a piece of ceramic, called a ceramic core, was inserted into the wax pattern at the time the pattern was formed it would be possible to reliably form the inside of the mold to ensure a high-quality casting.

In the 1950s the shape of these blades was relatively simple. Turbine blades are similar to miniature airplane wings. Pressing could produce ceramic parts for the cores, but the required tolerances could not be met easily. Other processes, such as casting, transfer molding, and injection molding, were explored to provide the technical solution. The casting process, sometimes called the Shaw process, consists of forming slurry of a material, such as fused silica, with an ethyl silicate binder. A gel agent, such as magnesia or ammonium hydroxide, is added, and the slurry is poured into a mold. As soon as the slurry gels, the mold is opened and the part is lit with a torch. The slurry is an alcohol base so that the burning immediately sets the shape prior to firing. The process is quite slow and difficult to control. In the injection-molding process, a slurry is prepared from a mixture of ceramic particles and wax and plastic binders. The slurry is warm, and it is injected into a cold metal die as in the injection-molding of plastics. After injection into the die, the green part is formed when the wax, plastic, and ceramic mixture cools. The green piece is removed, and the binder material slowly fired away. Finally, upon heating to a high temperature, the particles of ceramic sinter together. Transfer molding is a very similar process except that a thermoset silicone resin material is mixed with the ceramic particles. When the mixture is forced into a warm die, an irreversible setting occurs, and, after removal from the die, the green piece is fired to remove the organic portion of the binder.

Ceramic cores are usually formed from fused silica, because this material is not reactive with most of the alloys used in the investment casting of turbine blades. Furthermore, the ceramic must be removed from the casting, and fused silica is soluble in a strong base, such as sodium or potassium hydroxide. A fired ceramic core is highly porous, usually 25 vol% porosity, and this property is helpful in the chemical process that is used to clean the inside of the casting.

A major reason for the high efficiency of the modern jet engine is the very elaborate cooling schemes on the inside of the turbine blade casting. These designs are possible by the use of dimensionally precise ceramic cores that can be reliably incorporated into the investment casting mold. Ceramic cores must be able to be heated to the melting point of the metal alloys in the metal-casting process, and these ceramic pieces must retain an exact shape. The development of this technology has enabled jet engines to operate at turbine temperatures above the melting point of the metal alloys that are used to form the hot section hardware.

Abrasives, Cutting Tools, and Wear-Resistant Materials

Advances in Abrasive Materials Technology in the Past Hundred Years

by David A. Sheldon, Saint-Gobain Abrasives

Significant commercialization of abrasive products began with the switch from naturally occurring materials (quartz, emery, garnet, corundum, etc.) to synthetically produced materials. This began with E. G. Acheson's discovery of silicon carbide while attempting to produce diamond with electric-arc heating. He named this new material carborundum, and its commercialization led to the founding of the Carborundum Company. Within a few years B. Jacobs, C. M. Hall, and A. C. Higgins completed the first successful fusion experiments to produce synthetic corundum or fused alumina from bauxite. Norton Company's introduction of monocrystalline 32ALUNDUM in the mid-1930s was the most notable innovation in a series of advances associated with furnacing, raw-material purity, and use of additives to provide various microstructures, fracture characteristics, and grinding behaviors of fused-alumina abrasives.

Another innovation was the development of fused, solid-solution zirconia–alumina abrasives in the 1960s and 1970s by Norton. The 25%

zirconia abrasives (ZS and AF Alundums) provided toughness and durability in steel conditioning wheels, and the 40% zirconia (eutectic) material (NZ Alundum) provided both toughness and sharpness for cutoff and portable wheels and for coated abrasive products.

Sintered alumina abrasives were first introduced by Norton (76Alundum) in the early 1960s for use in the conditioning of stainless steel. This was produced by extrusion of finely ground bauxite and sintering in a rotary kiln at 1400°–1600°C. The most recent innovation occurred in the mid-1980s with 3M (Cubitron) and Norton (Seeded-Gel and Targa) introducing high-purity, microcrystalline, sol–gel-derived α-alumina abrasives. These abrasives represent current state-of-the-art technology relative to sharpness and durability in conventional coated abrasives and bonded (resin and vitrified) wheels.

The class of materials known as superabrasives began with General Electric's invention of synthetic diamond in 1955. This innovation removed the occasional shortages of natural diamonds and eliminated the variability in quality associated with minerals. In the late 1960s GE developed cubic boron nitride (Borazon). These materials are the hardest and most efficient (and expensive) for grinding both non-ferrous (diamond) and ferrous (cubic boron nitride) work materials.

Glass bonds for vitrified bonded abrasive products have seen significant innovations over the past 100 years. Upon the commercialization of silicon carbide and fused corundum, early grinding wheels were cast (puddle process) in potteries utilizing glaze raw materials (clay, feldspar, etc.) to form $Na_2O/K_2O–CaO/MgO–Al_2O_3–SiO_2$ glass bonds. In the 1920s, boron oxide was added to the bonds to reduce sensitivity to varying firing conditions and to eliminate devitrification. In the 1970s many manufacturers began to incorporate small amounts of lithium oxide in bonds to enhance mechanical properties (strength and durability) of wheels. Other glass technologies, such as glass-ceramics (controlled nucleation and crystal growth) and sol–gel processing have been studied but with little or no commercial success.

Acheson Process for Silicon Carbide

by Neil N. Ault, Saint-Gobain Corporation

Dr. Edward Acheson was attempting to produce diamond when he heated clay and a carbon rod electrically and discovered that he had formed a few very hard crystals. Others had preceded him in such experiments but did nothing with their results. He recognized that he had formed a new material, and thought he had formed a compound of carbon and alumina (corundum), so he called the material "Carborundum." Unlike others, he went on to analyze the crystals and found that they were silicon carbide.

Recognizing the high hardness of the material, he proceeded to sell it initially to gem polishers and later as the abrasive in a ceramic bonded grinding wheel. His innovation was the discovery and analysis of silicon carbide and the development of a prototype furnace for its commercial production, the design of which continues in use today. He obtained a patent in 1893 covering the "silicide of carbon, SiC" and the method of making same by heating a mixture of clay or silica and carbon to form silicon carbide.

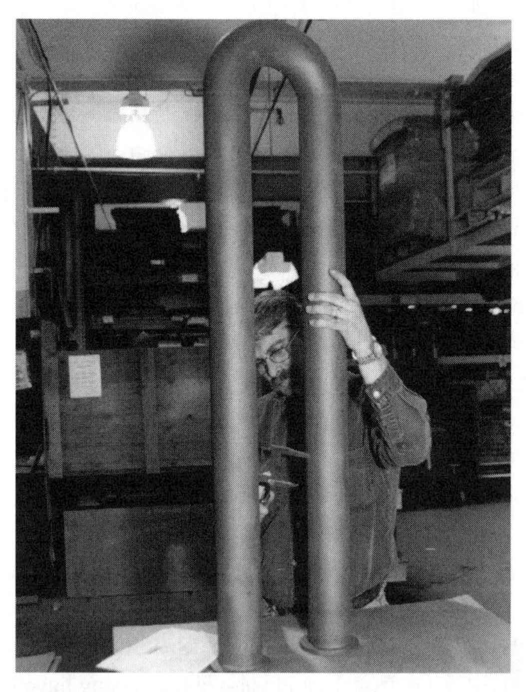

Figure 16. *Silicon carbide radiant "U" tube. Courtesy of High Performance Refractories, Saint-Gobain Industrial Ceramics, Inc.*

His furnace design uses a central resistive graphite core surrounded by a mixture of 60% sand and 40% coke. A typical furnace 45 ft long operates at a total energy input of about 54,000 kW·h, and it produces ~18,000 lb of silicon carbide.

Acheson established the Carborundum® company in 1891 to take advantage of this new material. By the end of the 19th century, he moved the manufacture of silicon carbide to Niagara Falls, NY, to take advantage of low-cost electric power.

The manufacture of silicon carbide grinding wheels began in the 1890s, but it was not until the 1900s that silicon carbide refractories were first produced, initially from low-quality crude and by 1923 from abrasive-grade crude and oxide bonds. Today, the worldwide capacity for silicon carbide grain is ~350 metric tons annually. About 35% of this is for abrasive applications, 15% for refractories, and 50% as metallurgical additions.

Responding to the problems of oxidation of oxide-bonded silicon carbide refractories, Swentzel of Carborundum in 1948 found that a mixture of silicon and silicon carbide could be nitrided to form a nitride-bonded silicon carbide. Another innovation by Washburn of Norton was silicon oxynitride-bonded silicon nitride in 1959. In 1931 Heyroth developed the first practical silicon carbide heating element with siliconized cold ends. Other important innovations for silicon carbide products were the slip-cast and sintered recrystallized silicon carbide by Fredriksson, now used for lightweight kiln furniture and furnace components for the electronics industry. Dense silicon carbide was developed by Prochazka by sintering (1973) and by Alliegro, Coffin, and Tinklepaugh by hot pressing (1956).

Synthetic Superhard Materials: Diamond and Cubic Boron Nitride

by Robert C. DeVries, General Electric Company, retired

In 1951 the General Electric Company undertook the search for a process for making diamond, and in 1955 announced the reproducible synthesis of abrasive grain, which was marketed as an industrial abrasive in 1957. After a long and controversial history of attempts to synthesize diamond, this reproducible synthesis of diamond using a metal solvent/catalyst at high pressures and high temperatures (HPHT) was a significant breakthrough and innovation in the production of superhard materials. Although the first product was a dirty, unattractive, friable abrasive, it proved to be especially effective in grinding carbide (which had been the motivation for the project in the first place). Furthermore, there was no natural product with a similar combination of properties, and continued progress in controlling shape, size, and toughness produced several other grades that met different needs of the grinding, polishing, cutting, and machining trades. Thus, synthesized diamond began to replace the natural form because of the ability to tailor the grain to specific needs, and now ~90% of industrial diamond is synthetic. The sintering of these diamond grains into a polycrystalline ceramic with diamond–diamond bonding followed from subsequent R&D under HPTP conditions, and truly remarkable cutting tools, drill bits, and wear parts took their place in the marketplace after 1973.

Cubic boron nitride (Borazon), which does not exist in nature, was invented and produced from HPTP processes by GE in 1957. This second-hardest material took its place in the marketplace first as abrasive grain and later as sintered ceramics. The unique chemical stability of this material made possible the cutting and grinding of iron, nickel, and cobalt-based superalloys that react unfavorably with diamond in these procedures.

In contrast to the HPHT processes, in the late 1960s and early 1970s, Russian and American scientists were experimenting with synthesis of diamond at low pressures (<1 atm) from the gas phase by chemical vapor deposition (CVD). The Russians were the first truly successful producers of diamond by this process, but the transition to the marketplace was primarily stimulated by Japanese scientists in the early 1980s. After this the effort was largely worldwide, and now it is possible to coat carbide and other materials with a polycrystalline diamond layer that rivals the HPHT-sintered material. The CVD process enables production of shapes and sizes not possible from the HPHT process, such as transparent, polycrystalline diamond on the order of 1 mm thick by 50 mm diameter useful for optical windows. The CVD process is as yet unlikely to challenge the HPHT process for cost and throughput of abrasive grain, however.

Cemented Carbide

by Henri Pastor, CERMEP

Throughout the 20th century there has been a constant improvement of the productivity of manufacturing industries as a consequence of the outstanding increase of the cutting-tool speed, itself linked to the outstanding development of cutting-tool materials. With global sales at $18 billion in 1994 and a forecast at $25 billion in 2005, the cutting-tool industry will enjoy solid growth in the next eight year period. The set of used materials is wide: high-speed steels, cermets, ceramics, superhard materials, and, predominantly, cemented carbides, which are basically tungsten carbide–cobalt (WC–Co) cermets. Cemented carbides consume ~55% of the world tungsten production, representing ~28,000 tons of tungsten. They take part directly (handyman's tools, ball pens, tire-studs, etc.) or indirectly (metal and wood machining tools, rock-drilling tools, public-works tools, wear parts, etc.) in our daily life.

WC was discovered 100 years ago by the French chemist and Nobel prize winner Henri Moissan. The first cutting tools were patented in 1923 by Karl Schröter (Osram Studiengescellschaft, Berlin) and marketed in 1927 by the Krupp Company (WIDIA™, from "wie diamant"). The first developed grades were made from hard but brittle WC powder grains (0.5–5 μm) cemented by relatively soft but tough cobalt metal (5–12 wt%) and were prepared by liquid-phase sintering. The properties depend mainly on the cobalt content and the WC grain size. The toughness of cemented carbide increases with the cobalt content; consequently its hardness and wear resistance decrease. Using coarse WC grains is beneficial for shock resistance, but, at a given cobalt content, the hardness is lower compared to fine-grained material. The hardness of binary grades varies between 80 and 2000 HV. Nowadays, the production of cemented carbides comprises the following steps: mixing of the constituents by ball or attritor milling, granulation by spray drying, die compaction, liquid-phase sintering, and, if necessary, not isostatic pressing (HIPing) or sinter HIPing.

In 1928 Krupp Company gave the General Electric Company an exclusive license to manufacture the WIDIA carbides in the United States, and General Electric then licensed Firth Sterling (Diamondite, Firthite, Firthalloy), Fansteel (Ramet), Callite Tungsten (Teco), and North American Cutting Alloys (Cutinit). The first significant development occurred in 1929 and consisted of replacing all or part of WC by other refractory carbides, particularly TiC, NbC, TaC, MO_2C. This was the start of cutting tools for the high-speed machining of steel In 1938 P. M. McKenna introduced Kennametal compositions (WC–TiC–Co), which are produced by crystallizing the carbide solid-solution (W,Ti)C out of a molten nickel bath (menstruum process).

Afterward came technical and technological refinements suggested or supported by a deeper scientific knowledge of the microstructure and the properties of these materials. The most important features during the past 30 years are:

- The introduction in 1969 by Sandvik AB of the first TiC-coated grades. This technology (CVD, PVD, etc.) of the more-and-more complex and functional coatings makes use of various hard materials, such as TiC, TiN, Ti(C,N), Al_2O_3, TiB_2, (TiAl)N, diamond, and cubic-BN.
- The introduction of titanium carbonitride-based cermets in the early 1970s.
- The development of Sialons and composite ceramics (Si_3N_4–TiC, Al_2O_3–SiC, Al_2O_3–TiC, Al_2O_3–ZrO_2, etc.) in the period 1970–1980.
- The sintering of diamond and cubic-BN in the early 1980s.

Nanophase Cemented Carbides

by Bernard H. Kear and Larry E. McCandlish, Rutgers University

Cemented carbides first attracted widespread commercial interest in the 1950s. Although at that time many different systems seemed promising, today the dominant position in the marketplace is held by WC/Co-based materials. Exceptional properties are realized when the constituent WC and cobalt phases are interconnected in three dimensions, forming a so-called bicontinuous structure. Such a structure displays high hardness, superior wear resistance, and good fracture toughness. Applications for WC/Co-based materials abound and include machine tools, drill bits, and wear parts.

Typically, WC/Co products are produced by mechanically mixing powders of the constituent phases, followed by cold pressing and liquid-phase sintering. The difficulty of uniformly mixing ultrafine WC and cobalt powders by mechanical means has heretofore limited the scale of the WC grain size attainable in the final sintered product to ~0.3 μm. During the past five years, Rutgers University and Nanodyne, Inc., have developed an alternative chemical processing technology that produces premixed powders at the submicrometer (nanoscale) level. In this process, spray drying is used to produce a homogeneous precursor powder of mixed tungsten and cobalt salts, followed by fluid-bed thermochemical conversion (pyrolysis, reduction, and carburization) to transform the precursor powder into the desired nanophase WC/Co product powder. Typically, the WC particle size in the agglomerated powder product is ~30–40 nm.

Conventional WC/Co powder consolidation technology, as practiced by many tool-making companies, involves liquid-phase sintering of powder compacts. When nanophase WC/Co powder is used as the starting material, some modifications to established sintering practice have been necessary because of the much higher surface area and more rapid sintering kinetics of the nanocomposite material. Moreover, to mitigate WC particle coarsening during sintering, a small amount of grain growth inhibitor (e.g., 0.2–0.8 wt% VC) must be added to the starting powder. When this procedure is used, fully sintered products can be produced without inducing significant coarsening of the WC grains. Compared with conventional micrograined material, fully sintered nanograined material displays superior properties and performance, which is due to its uniform ultrafine structure.

Using a pilot-scale production unit, Nanodyne is currently producing tonnage quantities of nanophase WC/Co-based powders, with compositions extending over the range of commercial interest from 3–30 wt% cobalt. A new production plant will come on-line soon, with a capacity of ~400 ton/year. Several established hard-metal companies are making use of these powders in the manufacture of sintered parts. Interest also is growing in the use of nanopowders as feedstock materials in thermal spraying of wear-resistant coatings. Nanodyne's chemical processing technology permits recycling of spent parts. This should lead to a further reduction in powder production costs, thus enabling an even broader range of applications for the future.

Monolithic Silicon Carbide by CVD

by Michael A. Pickering, Morton Advanced Materials

For many years, silicon carbide has been recognized as a good material for applications where superior attributes—such as hardness and stiffness, strength at elevated temperatures, high thermal conductivity, low coefficient of thermal expansion, and resistance to wear and abrasion—are of primary value. Silicon carbide has been produced as reaction-bonded, hot-pressed, or sintered. These forms of silicon carbide are usually not pure silicon carbide but contain additives or a second phase and in many cases are not theoretically dense. These forms have lower thermal conductivity and limited high-temperature oxidation resistance or high strength, which has limited their use in applications requiring temperatures >1000°C.

Since the turn of the century, chemical vapor deposition (CVD)—the reaction of gaseous chemicals in a chamber or furnace to form solid crystalline materials—has been used primarily for producing coatings on substrates or free-standing (<1 mm) wafers. In 1980, Morton developed a revolutionary, large-scale CVD process for producing infrared-transmitting materials up to 60 in. diameter and 4 in. thick in near-net-shape configurations. In the 1990s Morton turned to the development of CVD engineered ceramic materials that today include commercial production of theoretically dense, polycrystalline cubic β CVD Silicon Carbide® that is free of microcracks or voids.

The Morton CVD process uses precise control of temperature, pressure, reactant flows, and vessel geometry in a highly complex chemical environment. The reactant gases are metered into the deposition zone of the furnace under low-pressure conditions. As the gases contact the hot wall, they chemically react to form CVD silicon carbide that is grown molecule by molecule on shaped mandrels that form the wall of the furnace. Once the desired thickness is reached, gas flows are shut off and the furnace is brought back to room temperature and atmospheric pressure. The CVD silicon carbide replicates the shape of the mandrels. This allows for the production of flat plates or complex near-net-shape geometries. The production of near-net-shape parts reduces machining time and costs.

Solid, high-purity (>99.9995%) CVD silicon carbide has superior thermal, physical, mechanical, and optical properties when compared with silicon carbide produced by other methods. Components made of CVD silicon carbide are free of graphite and other sources of contamination.

CVD silicon carbide has proved superior in semiconductor, wear, and optical applications as well as other applications that require high temperatures (>1000°C) with long life cycles. Specific applications include rings, susceptors, and chamber components for 200 and 300 mm wafer processing. Wear applications include high-performance mechanical seals, mold blanks for direct molding of glass lenses, and high-performance bearings. CVD silicon carbide also is a lightweight reflective optics material that is used to fabricate mirrors, high-energy lasers, laser radar systems, and synchrotron X-ray optics.

SiC-Whisker-Reinforced Ceramics

by Terry Tiegs, Oak Ridge National Laboratory

Advanced Composite Materials Corporation (ACMC) developed a rice-hull-based production capability to manufacture SiC whiskers primarily to reinforce metal-matrix composites. SiC whiskers used for reinforcement are discontinuous, rod-, or needle-shaped fibers in the size range of 0.1–1 μm in diameter and 5–100 μm in length. Because they are nearly single crystals, the whiskers typically have very high tensile strength (up to 7 GPa) and elastic modulus (up to 550 GPa). Collaborative work on hot pressing led to fabrication of the first whisker-reinforced Al_2O_3 composites at Oak Ridge National Laboratories (ORNL) in 1982. The initial results showed such promise that large-scale development programs were started to exploit the materials, and several years of research followed.

The mechanical property improvements observed with the incorporation of SiC whiskers into ceramic matrices were unprecedented. For example, the fracture toughness of Al_2O_3 was increased from ~3.0 to 8.5 MPa·m with the addition of 20 vol% whiskers. This was accompanied by fracture strengths of 700–800 versus ~400 MPa in unreinforced Al_2O_3. Just as importantly, these property improvements were retained to elevated temperatures, unlike some other toughened ceramic systems. Remarkably improved thermal shock and creep resistance also were observed. Subsequent studies have examined the SiC-whisker reinforcement of numerous ceramic-matrix systems, including mullite, ZrO_2, glass, spinel, cordierite, Si_3N_4, B_4C, and combinations of these materials.

Research into the toughening behavior responsible in the composite materials shows that crack bridging, whisker pullout, and crack deflection are the major toughening mechanisms. One of the keys to the behavior of SiC-whisker-reinforced composites is that, for this mechanism to operate, debonding along the crack–whisker interface must occur during crack propagation and allow the whiskers to bridge the crack in its wake.

During the same time period that the early research efforts were being done after the initial laboratory results, product development efforts were being conducted at ACMC. This work led to the use of SiC-whisker–Al_2O_3 composites for cutting-tool applications. The first commercial cutting tools based on this technology were introduced in the

United States by Greenleaf Corporation in April 1985. ACMC and Greenleaf received several awards for the pioneering development of this cutting tool, including an *R&D Magazine* IR-100 Award and the Corporate Technical Achievement Award from The American Ceramic Society. At the present time, Sandvik AB of Sweden also markets a SiC-whisker-reinforced cutting tool in other parts of the world. In the area of cutting tools, it was found that the addition of SiC whiskers to Al_2O_3 improved the strength and fracture toughness, and did not compromise the hot hardness of the matrix. This combination revolutionized machining of high-nickel alloys that are used in the jet-engine industry. The SiC-whisker-reinforced Al_2O_3 enabled a tenfold increase in metal removal rates at increased speeds. For example, in one reported case history, changing from a conventional tool to a SiC-whisker–Al_2O_3 tool reduced a 5 h machining operation of inconel to 20 min. In addition to cutting tools, SiC-whisker-reinforced composites are used in a wide variety of applications in can forming punches, extrusion dies, and other wear environments.

5

Functional Use of Ceramics

The area of electromagnetic and mechanical applications of ceramics, including combined applications, has experienced many innovations in the past century. The special electromagnetic properties of ceramics have made many new technologies possible, especially electronic and optical technologies.

Innovations in ceramic dielectrics based on barium titanate and related materials have led to multilayer ceramic capacitors and to microwave ceramic components in filters that make possible cellular telephones. Zinc oxide varistors provide protection for devices ranging from electronics to transmission lines. The discovery of ceramic superconductors with critical temperatures above the boiling point of liquid nitrogen seems likely to lead to many applications.

Piezoelectric ceramics have been the enabling technology for applications as diverse as submarine detection and medical ultrasonic imaging.

The discovery of ceramic ferrites in the 1930s has led to many magnetic applications of ceramics. High-frequency power supplies, transformers, and magnetic read/write heads are made in the billions per year. Other major applications are in magnetic recording on tapes and disks and in video recorders.

The phenomenon of transparency in glass and certain crystalline ceramics has led to many special optical applications. Uncooled infrared detectors are used in night vision devices, cameras, and missile guidance systems. Polycrystalline electrooptic ceramics are used for eye protection, in optical data recording and processing, and in image enhancement systems.

An important field that is totally new since the 1960s is that of glass and ceramic lasers. The development of low-loss optical fiber has revolutionized high-capacity communications. Light transmitted through such fibers must be modulated, amplified, and filtered. The invention of the erbium-doped fiber-optic amplifier allows amplification of the optical signal directly and is replacing the older process of transforming the optical signal to an electronic signal, amplifying it electronically, and reconverting it to an electronic signal. Low-loss optical filters are based upon ultraviolet treatment of photosensitive glass.

Sensors are a major area of application for ceramics and glass. A relatively simple example is the positive temperature coefficient resistors used as temperature sensors. An important class of materials functions both as a sensor and an actuator. These materials include piezoelectric and electrostrictive ceramics as well as magnetostrictive and shape-memory alloys. Another type of sensor is the fiber-optic sensor whose applications include measurement of strain, pressure, vibration, fluid flow, temperature, voltage, current, and rotation rate as well as biomedical and chemical recognition. The zirconia oxygen sensor is an essential element of all modern automobile engines.

Ceramics for structural (i.e., load-bearing) applications have been greatly improved by many innovations in the past century. Ceramic armor was developed in the 1950s to give improved protection at practical weight and cost. The development of continuous ceramic fibers led to new, tough ceramic-matrix composites used in defense and aerospace applications. Melt infiltration of silicon carbide has produced useful ceramics and composites. The discovery of transformation toughening in zirconia-based ceramics has produced tough and strong ceramics. Concrete based upon portland cement has seen remarkable improvements. Structures using the new reactive powder concrete are beginning to appear. Chemical tempering of glass is an older development but one that remains of great practical importance.

Strong ceramics are beginning to find practical use. *In-situ* reinforced silicon nitride appears to be a particularly promising material. Successful special applications of silicon nitride include turbocharger rotors in certain automobiles and as bearings for liquid-oxygen pumps in the space shuttle engines.

Electrical Insulators, Dielectrics, and Conductors

Barium Titanate

by Hans Thurnauer, IESC, retired

During the era of Edison, 1876–1892, ceramic dielectrics played an important role ranging from insulators on high-voltage transmission lines to low-voltage electrical applications in industry and homes. The preferred material was feldspatic porcelain, similar to the one used for sanitaryware.

With the arrival of the electronic age new materials with low dielectric loss had to be developed to meet the requirements of high-frequency electronic circuits, including high voltages. Steatite (magnesium aluminum silicate) replaced porcelain for general insulation. Another material, rutile (TiO_2), was successfully used in capacitors, in the form of disks and tubes.

After several years of active service, the need for even higher dielectric constant (K) values became apparent. Barium titanate ($BaO{\cdot}TiO_2$) was a likely candidate, with a K value at least a decade higher that that of rutile, but only within a narrow range of temperature and voltage. The original investigators of barium titanate dielectrics were Wainer and Salomon in 1942, under a secret U.S. government contract; secrecy was lifted in 1945. Research with additives and use of ultrafine crystal sizes led to acceptable compounds with K values of ~1200. Ceramic capacitors using polycrystalline barium titanate played an important role in military applications, such as guided missiles (sidewinder) in the Korean war and are today in widespread use.

A better understanding of the effects of microstructure and additions, called "shifters" (of Curie temperature), resulted in K values ranging from ~4000 to 7000. These compounds have replaced the original K1200 material.

Barium titanate is a synthetic mineral; i.e., it is not found in nature. It can be made in various ways, of which the most common is solid-state reaction of barium carbonate with titanium dioxide calcined at high temperatures. Other routes are wet processes based on soluble barium and titanium compounds coprecipitated and calcined.

Figure 17. *Multilayer ceramic capacitors for surface mounting. Courtesy AVX Corporation.*

By far the largest use of barium titanate is for capacitors. The present consumption is estimated to be 14 $\times$ 10^6 lb/year, which is equivalent to ~30 billion capacitors in the form of disks and multilayer ceramic capacitors (MLCC). The latter consist of up to 250 layers of dielectric, ~4–7 μm thick, coated with metallic electrodes. The process of manufacture is a high technology achievement that permits almost limitless capacities per unit.

Piezoelectricity of certain crystals was discovered in 1880 by Pierre and Jacques Curie. Later, in the 1940s, microcrystalline barium titanate ceramics were classified as ferroelectrics. One characteristic is piezoelectricity. Barium titanate is not piezoelectric per se. It becomes so when heated and cooled below the Curie temperature under the influence of an electric field that orients the domains into parallel alignment of its dipoles. It is important that the crystal structure of piezoelectrics is coarse, ~10 μm, in contrast to the size of ~1 μm in capacitors. The main advantage of polycrystalline piezoceramics over single crystals is that they can be made in sizes and shapes unavailable in single crystals.

The first application of piezo barium titanate was in World War II for sonar transducers under water to locate submarines. Among peacetime applications are ultrasonic sound generators, measuring instruments, and ultrasonic cleaners.

Microwave Dielectric Ceramics

by Henry O'Bryan, Bell Laboratories

Electronic ceramics find applications in microwave communication as ferrites in circulators, garnets in phase shifters, and dielectrics and insulators both as substrates on which circuits are fabricated and as elements in microwave integrated circuits. Although the application of dielectric ceramics as resonators for frequency selection was conceived about 1940, their use required materials with a unique combination of moderate dielectric constant, low microwave loss (high Q), and temperature stability. TiO_2 with a very large temperature coefficient of resonant frequency (TCF) of +400 ppm/°C was not practical. Filter needs were therefore met by air-filled copper cavities that were large and had a TCF value of ~17 ppm/°C. When greater temperature stability was needed, an expensive Invar cavity was used. Filter dimensions depend on the frequency of use and are proportional to $K^{-1/2}$ where K is the dielectric constant. Thus a suitable dielectric also promised an impressive reduction in filter size.

In 1969 Readey and co-workers at Raytheon developed $BaTi_4O_9$ with low loss, K = 36, and TCF = 25 ppm/°C. Shortly thereafter a Bell Laboratories group—O'Bryan, Plourde, and Thomson—reworked the Ba-Ti-O system and found an overlooked compound ($Ba_2Ta_9O_{20}$) that had equivalent low loss but slightly higher K and a TCF as low as Invar (TCF = 2 ppm/°C). When incorporated into filters, resonators of this compound increased capacity of the channel microwave radio system, which at that time carried the bulk of long-distance telecommunication. In addition there was a large cost advantage over Invar and size advantage over any metal cavity filter. Later this ceramic was used in the combining filter of the base stations for the first generation of the cellular phone system.

Since then other microwave dielectric ceramics have been discovered, and temperature-stable materials with K = 12 to 90 are now available. The ceramic systems most commonly used for microwave resonators are $Ba_2Ti_9O_{20}$, $(Sn,Zr)TiO_4$, $Nd_2BaTi_4O_{12}$, and $Ba(Mg,Ta)O_3$. New filter design and manufacturing methods have placed microwave dielectric ceramics as resonant elements in cellular and PCS hand sets and base stations, and GSM systems.

One interesting filter design is based on the coaxial resonator. The ceramic is in the form of a tube-shaped object, and most of the surfaces (including the center hole) except for one end face of the tube are covered with high-conductive metal. This forms a quarterwave resonator that resonates at the frequency of $c/(4LK)$, where c is the speed of light, L the length of the tube, and K the dielectric constant of the ceramic. The configuration of dielectric ceramic filters has gone through many phases over the years. The filter was originally made by linking-up several single quarterwave resonators and then soldering discrete capacitors or inductors in between the resonators to serve as coupling elements. Later on, monolithic block filters were made by forming a single slab with several parallel holes in it. The slab was coated with metal on all surfaces except one of the two surfaces with holes. On the un-metallized surface, coil inductors or interdigital capacitors were screen printed to form the coupling elements. More recently, the monolithic block filters have been designed with three-dimensional features on one side of the block, and after metallization these three-dimensional features function as the coupling elements.

The applications of dielectric ceramic filters are likely to expand in the future, especially for the applications in the 2–5 GHz range where other filter technologies cannot meet the performance and cost requirements. The success of the dielectric filter technology will continue to depend on the development of new dielectric ceramics, innovative filter designs, and precision manufacturing processes. Several hundred million dielectric filters and resonators are made each year. It can be said that dielectric resonators are the enabling technology for miniaturizing equipment for all wireless telecommunications.

Ceramic Ion Conductors

by Ronald S. Gordon

Ceramic materials, such as α-alumina (corundum), are used extensively as electrical insulators in such applications as automotive spark plugs. Many of these materials have very high electrical resistance, corrosion resistance, and high-temperature characteristics. Refractory and corrosion-resistant ceramic materials also can be made into electronic semiconductors and ion-selective ionic conductors with attractive performance characteristics as either electrodes or electrolytes in energy storage batteries, power-generating fuel cells, thermoelectric power generators, chemical sensors, oxygen pumps, and other electrochemical devices. A combination of corrosion-resistant, refractory, and electrically conductive properties is now making possible the commercialization of two principal ceramic materials (β´´-alumina and cation-stabilized zirconia) as ion-conducting electrolytes in the sodium–sulfur and ZEBRA batteries and in the solid-oxide fuel cell. Beginning in the mid-1960s, these systems have been under commercial development worldwide for applications in electric vehicles, electric power generation plants, and load-leveling operations in electric utilities. Full-scale commercialization of these technologies is expected early in the next century.

β-alumina and, subsequently, β´´-alumina were discovered to be excellent conductors of sodium ions by Weber and Kummer in the early 1960s. Zirconia-based oxides were found to exhibit high oxygen-ion conductivity at elevated temperatures by Nernst in 1900. In 1943 Wagner proposed that oxygen-ion vacancies were the mobile species carrying the ionic current in these ceramic materials. Wagner further proposed that these vacancies were generated on the anion sublattice by incorporating oxides of aliovalent cations (e.g., Ca^{2+} and Y^{3+}) into the cubic crystal structure.

Polycrystalline β´´-alumina, the preferred, commercial ceramic with an optimum sodium-ion conductivity, is the ion-selective solid-electrolyte membrane in the sodium–sulfur and the ZEBRA batteries, which both have a liquid-sodium anode and either a molten-sulfur or nickel choloride-based cathode. These batteries, with attractive energy densities, operate at temperatures of ~300°C. Sodium ions carry the current through the ceramic electrolyte during discharge and charge of the battery. Yttria-stabilized zirconia is the ion-selective membrane in the solid-oxide fuel cell. The fuel cell, which is projected to operate at temperatures between 600° and 1000°C, continuously electrochemically oxidizes a fuel, such as hydrogen, carbon monoxide, or methane. At the temperature of operation, yttria-stabilized zirconia has the requisite oxygen-ion conductivity for efficient transport of oxygen ions through the ceramic electrolyte in the electrochemical fuel cell. This results in the economical generation of electric power in an environmentally clean manner.

β-alumina is actually misnamed. It is not a polymorphic form of aluminum oxide, as its name would imply. It is a sodium aluminate compound with a crystal structure consisting of spinel-like blocks of aluminum oxide separated by two-dimensional conduction planes in which mobile sodium ions reside. At the time of its original discovery in 1916 by Rankin and Mervin, the presence of sodium oxide (11–16 mol%) in the crystal structure was undetected. The optimum composition with the highest sodium-ion conductivity in a commercial polycrystalline ceramic is $Na_{1.67}Li_{0.33}Al_{10.67}O_{17}$ (or ~13.2 mol% Na_2O and ~2.6 mol% Li_2O). The lithium oxide stabilizer serves two functions: (1) stabilization of the β´´ crystalline phase and (2) enhancement of the sodium-ion conductivity. Lithium cations substitute directly for aluminum ions in the spinel blocks and are electrically compensated by additional mobile sodium ions in the conduction planes. Because of a low activation energy for sodium-ion conduction and its characteristically high ionic conductivity (i.e., ~0.25 $(\Omega \cdot cm)^{-1}$) at the relatively low temperature of 300°C, β´´-alumina is considered to be a fast-ion conductor. It is one of the few such conductors close to commercial exploitation that can be fabricated into practical polycrystalline ceramic forms with suitable electrical and mechanical properties.

Zirconia can be stabilized in the fluorite-cubic crystal structure at elevated temperatures by the addition of ~8 mol% yttria. The yttrium ions substitute directly on the cation lattice for the tetravalent zirconium ions. Electrical charge compensation is accomplished by the formation of compensating vacancies on the oxygen-ion sublattice. It is the presence of these oxygen vacancies that results in the oxygen-ion conductivity in this material. Yttria-stabilized zirconia is currently the electrolyte of choice in the solid-oxide fuel cell with an ionic conductivity of ~0.1 $(\Omega \cdot cm)^{-1}$ at 1000°C. This material has the requisite ionic conductivity and is thermodynamically stable in the required crystal structure in the temperature range (i.e., 600°–1000°C) projected for the operation of the solid-oxide fuel cell.

Superconducting Ceramics

by Robert J. Cava, Princeton University

Superconducting materials conduct electricity with exactly zero energy loss. This occurs below an electronic phase transition at temperatures very close to absolute zero. Superconductors can be used, for example, to make the world's strongest electromagnets, because the high electrical currents needed to generate high magnetic fields pass through the coils without losing energy. Conventional metallic alloy superconductors are used to generate the magnetic fields in today's MRI machines. In 1986 basic and applied science were turned upside down by the discovery of superconductivity at unprecedentedly high temperatures in ceramic materials based on copper oxide. Although ceramics are well-known for their applications in a variety of important electronics technologies, such uses are generally based on the fact that ceramic materials are excellent electrical insulators—with virtually no-one ever imagining that ceramic materials might also include the best electrical conductors ever discovered.

Superconductivity was first observed in 1911 in elemental mercury cooled to very low temperatures by using the newly produced coolant liquid helium (stable at 4 degrees above absolute zero, i.e., 4 K). A variety of pure-metal elements were found to be superconducting below a critical temperature (T_c) of a few degrees K. Their potential for producing powerful magnets was immediately realized, but the superconductivity was found to be easily destroyed by the magnetic fields generated by the electromagnets. After several decades of research, the 1950s saw the discovery of new types of superconductors, niobium-based metal alloys, which maintained their superconductivity in the presence of high magnetic fields. The highest T_c, 23 K, was found for the metal alloy Nb_3Ge. Although still a very low temperature, the development of superconductors' applications began. In addition, a theoretical explanation of superconductivity was found in 1957 (the BCS theory). For the metallic-alloy superconductors, the mechanism involved the coupling of the conducting electrons to each other in a collective high-energy state through interactions with the underlying crystal lattice.

For the next three decades, scientific and technological progress in superconductivity proceeded at a normal pace, without the appearance of any new materials whose T_c or current-carrying capability exceeded those of the niobium-based alloys of the 1950s. A few ceramic superconductors, such as barium lead bismuth oxide and lithium titanium oxide, were known, but they were regarded almost universally as unremarkable. Two Swiss scientists, J. G. Bednorz and K. A. Muller, thought otherwise, and began looking in conducting ceramic oxides for superconductivity, with the success of their two year search revealed in late 1986 with the announcement of superconductivity at 28 K in lanthanum barium copper oxide. Some researchers immediately realized that the new ceramic superconductor represented something truly unexpected, and within a few months, another copper oxide-based ceramic was found, $YBa_2Cu_3O_7$ (123), with the stunning T_c of 92 K, above the boiling point of liquid nitrogen, a simple to use, inexpensive coolant. In the years that followed, the discovery of new superconducting ceramics proceeded at a dizzying pace, with more than 50 types of copper oxide superconductors presently known. Several important families of materials followed the discovery of the 123 compound, such as the bismuth; thallium; and mercury-based copper oxides, including, for example, the very-high-T_c materials $Bi_2Sr_2Ca_2Cu_3O_{10}$ ($T_c \approx 110$ K), $Tl_2Ba_2Ca_2Cu_3O_{10}$, ($T_c \approx 125$ K), and $HgBa_2Ca_2Cu_3O_9$ ($T_c \approx 130$ K at ambient pressure and 160 K at high applied pressures, the highest T_c known).

The superconductivity in the high-T_c ceramic superconductors occurs on checkerboard-geometry planes of CuO_4 squares within their crystal structures. The remainder of the component atoms provide a framework for the planes, control the amount of electrical charge on the planes, and provide coupling between the planes so that current can flow in three dimensions. Experiments have shown that the character of the current carriers is such that no previously known theoretical models for electronic transport or superconductivity can apply to the ceramic superconductors, making their understanding one of the great unsolved problems in science. Their study has been one of the most fertile areas for basic and applied physical science research in the past decade, and many new physical phenomena have been found. Although the promise for applications is great, the fundamental nature of the superconductivity has made such applications not yet possible. For ceramics prepared by conventional sintering methods, atomic disorder at the grain boundaries disrupts the superconducting current flow, especially in the presence of a magnetic field, resulting in energy loss. Special melt processing that results in a highly oriented ceramic has been pursued as a solution to this problem, and progress is continuously being made toward the fabrication of long-length high-current-capacity cables, especially in the bismuth-based copper oxides. In an advanced development stage are filters for microwave communications applications, which use the ceramic superconductor 123 in resonant cavities. Such filters have exceedingly low losses in the microwave frequency range and may be the first viable commercial products made based on high-T_c superconductors.

Zinc Oxide Varistors

by Lionel M. Levinson, General Electric Company

Electrical and electronic systems and devices require protection against unwanted voltage surges. A variety of devices (spark gaps, selenium rectifiers, silicon diodes, and silicon carbide) have been used for this purpose, starting from the days of Thomas Edison.

The ability of electronic ceramic devices based on zinc oxide to provide protection against overvoltages was first discovered in the USSR in the 1950s, but its practical application was never realized. Zinc oxide varistors (variable resistors) were independently rediscovered in Japan in the 1960s and brought to commercial application in Japan and the United States in the 1970s for consumer electronics and electric utility power transmission and distribution.

Zinc oxide varistors operate by being able to change their resistance dramatically (more than 10 orders of magnitude!) when the voltage increases above a safe level. In practice, the varistor is connected in parallel with the device to be protected, and its "breakdown voltage" is chosen so that its electrical resistance is very high (insulating) when the voltage is normal or safe. When the voltage increases above the safe level because of abnormal conditions, the varistor "breaks down," i.e., begins to act as a conductor and shunts any abnormal voltage and current away from the device to be protected.

The extremely nonlinear electrical behavior of this device results from its ceramic structure, which comprises grains of conducting zinc oxide surrounded by highly insulating and very thin (a few nanometers) grain boundaries. Below breakdown the grain boundaries isolate the conducting grains and the material is insulating. When the voltage increases to a critical value, the high electrical field across the thin grain boundaries causes electrons to transfer across the grain boundaries and the material conducts.

Zinc oxide varistors are now the voltage surge protection device of choice for electrical systems with voltages ranging from tens of volts to megavolts. They are relatively inexpensive and generally have perform-

Figure 18. *Lightening arrester being connected to a high-voltage transmission line. The zinc oxide varistor material forms a stack tens of feet high inside the fluted porcelain arrester housing visible. Courtesy GE Co.*

ance superior to other voltage surge protective devices. Varistors have a fast response (less than a nanosecond), which is needed to protect modern electronics. Many billions of these devices are in use to protect TVs, computers, stereos, and homes from lightning-induced voltage surges. Being ceramic, zinc oxide varistors are rugged, can be fabricated in a wide variety of shapes and sizes, and can be made large enough to protect high-voltage high-energy power transmission systems. The figure shows a porcelain lightning arrester housing containing many hundreds of pounds of zinc oxide varistor material being connected to protect a high-voltage power transmission system.

Transducers and Actuators

Piezoelectric Ceramics in Medical Ultrasonic Imaging

by T. R.(Raj) Gururaja, Hewlett Packard Company

The transducer is the heart of any medical ultrasound imaging system. Currently, most of the transducers utilize lead zirconate titanate- (PZT-) based piezoelectric ceramics to convert electrical energy to mechanical energy, and conversely, convert mechanical energy to electrical energy.

The applications of medical ultrasonic imaging span a wide frequency range, from ~1.5 to 50 MHz, depending on the organs to be imaged. For abdominal, obstetrical, and cardiological applications, the frequencies range from 2 to 5 MHz. For pediatric and peripheral vascular applications, the range is from 5 to 10 MHz. For imaging small objects, such as the eye, and for many other emerging applications, such as intracardiac and intravascular imaging, frequencies range from 10 to 50 MHz.

The reason for the extensive application of PZT ceramics in ultrasonic imaging transducers is a combination of several critical characteristics: relative high electromechanical conversion constant (50%–75%, depending on composition and geometry) leading to higher sensitivity and imaging of deeper organs in the human body; relatively high dielectric constant with a large range of values (1000 to 5000) helps to optimize electrical matching to the electronic circuits of the system; stability of material properties, economy, ease of fabrication, and mass production; processed by conventional technology to build fine-scale transducer arrays.

In 1954 B. Jaffe, R. S. Roth, and S. Marzullo reported on the remarkable piezoelectric and dielectric properties of PZT-based piezoelectric ceramics. At the morphotropic phase boundary (MPB) between the tetragonal phase and rhombohedral phase, in the vicinity of Zr/Ti = 53/47, both the dielectric and piezoelectric constants exhibit a sharp maxima. Since then, the compositions near the MPB have been modified with various atomic substitutions to prepare PZT ceramics with a broad range of properties classified as soft and hard piezoelectrics. Advances in ceramic processing techniques, such as improved ceramic powder characteristics and sintering techniques, including hot isostatic pressing and hot pressing, have significantly reduced the porosity in PZT ceramics.

Initially the transducers consisted of a single element of PZT with a diameter of 1–2 cm. Currently the transducers are made of a linear or curved array of transducer elements. A single transducer could have anywhere from 32 to about 300 elements, each of them wired separately to be connected to electronic circuitry. The width of each element ranges from ~500 to <100 μm, depending on the aperture size and the frequency of use. The thickness of transducer elements range from ~700 μm for a 2 MHz transducer to ~30 mm for a 50 MHz transducer. The resolution of the ultrasound images depends on the wavelength of ultrasound, which is inversely proportional to the frequency. For example, a 10 MHz transducer can resolve structures as small as 200–300 μm.

Although material requirements (dielectric, piezoelectric, and mechanical) have changed significantly over the past two decades, PZT remains the most commonly used material in transducers for medical imaging. Furthermore, two-phase piezoelectric composite materials developed to tailor and optimize material properties also use PZT as the active material for transduction. Piezoelectric polymers composed of vinylidine fluoride and trifluoroethylene copolymer (P(VDF-TrFE)) have acoustic properties closer to the human body compared to that of PZT ceramics, but the coupling constant and dielectric constant are not high enough to replace PZT ceramics. Recent investigations show that single crystals of relaxor ferroelectrics, such as lead magnesium niobate–lead titanate (PMN-PT) and lead zinc niobate–lead titanate (PZN-PT) solid solutions have demonstrated a remarkably high coupling constant, $k_{33} \approx 90\%$, compared to 70%–75% in PZT ceramics. Investigations are underway to explore the utility of these materials in the next generation of devices.

Piezoelectric Ceramics from Igniters to Computers and Telecommunication Devices

by Koichi Kugimiya, Matsushita Electric Company

Titania is a very useful material. It is widely used for face powders. In the very early days of radio broadcasting, titania was used as a dielectric material, especially in microwave applications.

As electronic devices developed dramatically, ceramics have been continuously improved and made very complex in constituents as requirements have been made ever higher.

In 1954 B. Jaffe obtained compounds near the morphotropic phase boundary in the solid-solution system $PbZrO_3$–$PbTiO_3$, commonly called PZT, that have temperature stability and high piezoelectric effect. Subsequently various modifications of these materials have been made, such as alkaline-earth-ion substitution for lead and substitution of zirconium and titanium sites by tin or niobium. Superior properties were obtained that opened up a wide field of applications.

In 1960 Smolenskii introduced a new family of complex perovskites, $A(B',B'')O_3$ where B´ and B´´ ions are in different ionic states and compensate an ionic charge difference. Stimulated by this work, research groups at Matsushita Electric Industry, Co., Ltd., attempted to combine the above two technologies in the early 1960s and developed a series of complex solid-solution perovskite materials based on PZT in the $Pb(Mg_{1/3},Nb_{2/3})O_3$–$PbZrO_3$–$PbTiO_3$ system that is now called PCM. Because the morphotropic phase boundary forms a line in the ternary system, one can now choose a wide variety of combinations of dielectric and piezoelectric values that are most suitable for various applications. PCM ceramics are now used as igniters, ceramic filters, actuators, speakers, and other applications in radios, TVs, computers, telecommunication systems, and transducers for medical ultrasound.

Piezoceramic Vibration Control

by Brian Mulcahey and Ronald L. Spangler, Active Control eXperts,. Inc.

Vibration problems have traditionally been solved by mounting equipment on vibration isolation (rubber pads or foams) and using mechanical dampers. A new class of vibration control technology—based on piezoceramics—now provides another option. Piezo-based dampers are efficient and easily tuned to the exact frequency that needs damping. As a result, they often outperform conventional dampers in a more compact package.

A notable example is the use of piezo dampers on skis. Schussing skis are prone to vibration. The frequency of these vibrations vary depending upon surface conditions and speed. The most difficult conditions are high speed and icy surfaces, which often combine to create large oscillations in the ski. These vibrations lessen the contact area between the ski edge and snow surface, reducing stability and control. This makes skiing in general more dangerous, and reduces the maximum speed ski racers can travel.

To produce more stable skis, manufacturers have attempted viscoelastics, tuned mass dampers, and composite materials. All these materials are effective to some degree, but each has its drawbacks, from temperature sensitivity (viscoelastics) to weight (tuned mass dampers) to cost (composites).

K2 corporation (a manufacturer of skis) recognized the potential performance of "smart materials" to control ski vibrations. Working with Active Control eXperts (ACX) they have developed a successful piezo damper. The ACX piezo damper acts as an electronic shock absorber. The piezo damper first converts the mechanical energy of the vibration into a voltage. This voltage is then applied across a resistive "shunt" circuit, which dissipates the energy as heat. In this manner, the damper acts as an energy transducer, converting the mechanical energy to electrical energy to heat energy. The shunt circuit is tuned to the specific frequency of the first vibration mode of the ski (roughly 10 Hz to over 100 Hz, depending on the model). Therefore, by attacking the vibrations that most affect ski performance, the piezo damper can reduce vibrations by ~50%. Reduced vibration provides a smoother ride, more responsive turning, and greater stability at high speeds, letting the user ski in control over a wider variety of conditions. This translates to a safer, more versatile, and more fun ski.

The ACX QuickPack™ strain actuator used in ski vibration damping is designed for broad commercial applicability beyond skis—or even sports equipment in general—to include medical, consumer, industrial, automotive, and aerospace applications. This actuator packages piezoceramics in a protective skin with preattached electrical leads. It makes fragile piezoceramics much easier to work with and easier to integrate into OEM products. The benefits that this device offers over raw piezoceramic devices include no soldered leads that can break, mechanical protection of the piezoceramic elements, integration of electrical components, electrical isolation, and protection against environmental damage.

There are two basic types of piezo dampers: passive and active. Passive piezo dampers utilize either a resistance–capacitance (RC) shunt circuit or a resistance–inductance (RL) shunt circuit. Most skis with piezo dampers use an RC shunt circuit, because the target mode of vibration is a first bending mode. RL dampers are preferred for higher-end race skis because the target mode of vibration is a higher-order bending mode.

Active systems use piezo actuators as part of a closed-loop vibration-control system where sensors measure oscillation, and onboard electronics command the actuators to generate motions that cancel out the vibrations. Active systems are used to alleviate excessive vibrations in the twin tails of F-18 aircraft.

Magnetic Ceramics

Hard Ferrites

by Alex Goldman, Ferrite Technology Worldwide

Whereas soft ferrite components are wire-wound and perform alternating-current electrical functions, hard ferrites are permanent magnet materials and are used to provide a direct-current magnetic field. Although both ferrites are mixed oxides with iron oxide as the major constuent, the two crystal structures and chemistries, as expected, are quite different.The soft ferrites are basically cubic without a single preferred magnetic direction, but hard ferrites need this magnetic anisotropy and thus have a hexagonal crystal structure with a preferred *c*-axis.

Hard ferrites have the magnetoplumbite ($PbFe_{7.5}Mn_{3.5}Al_{0.5}Ti_{0.5}O_{19}$) crystal structure. In hard ferrites, the formula is $MFe_{12}O_{19}$ or $BaO{\cdot}6Fe_2O_3$ where M is barium, strontium, or lead. The elementary cell consists of 10 layers. There are four layers of four oxygen ions each followed by a fifth layer containing three oxygen ions and one barium ion. This pattern is repeated, but this time, the barium ion in the fifth layer is in the diametric position requiring the 10 layers for complete uniqueness. The iron ions that possess the required spin magnetic moments are located interstitially.

Hard ferrites were developed in the Philips Research laboratories by Went *et al.* in 1951. The magnetic moment of a hard ferrite suffers from two effects in comparison to the metallic magnet. First is the dilution of the magnetic ions (iron) by the large oxygen ions. Second is the antiparallel alignment of some of the iron spins in the hard ferrite as opposed to parallel alignment in metals. Therefore, hard ferrites have lower (¼ to ⅓) permanent magnetizations (density of net magnetic moments) compared to their metallic counterparts, such as Alnico 5, but they have much greater resistance to demagnetization (5–6 times). The former property is the remanence, B_r, while the latter is the coercive force, H_c. A figure-of-merit is the product of the two, called the maximum-energy product (B_rH_c)max; the products of the two types of materials are close. To compensate for the low B_r, hard ferrite magnets have a wide cross-sectional area and a short length. In Alnico 5, the magnet shape is long and thin. There are two varieties of hard ferrites, the nonoriented, or isotropic, with the hexagonal axes of the grains randomly aligned, and the oriented, or anisotropic, with grains oriented with their *c*-axes parallel, greatly increasing B_r. The latter is superior in energy product and is more costly; therefore, its use is restricted to more demanding applications. By variation of chemistry, additives, and processing, hard ferrites are available with particular ceramic properties that favor specific applications. Barium ferrite was the original hard ferrite, but strontium ferrite has recently become the material of choice. Aside from the sintered magnet, use of the plastic-bonded hard ferrite has gained in importance

In the processing, the iron oxide and the barium or strontium source (usually carbonates) are usually dry mixed. The Fe_2O_3:SrO ratio is commonly 5.5. The mixture is granulated in a pelletizer and prefired at ~1250°C. Excess strontium oxide and small additions of silica and boron oxide help control grain growth. After prefiring, the hard pellets are wet milled to a particle size smaller than critical domain size (<1 μm). For oriented material, the slurry is filter pressed in a magnetic field. The nonoriented material is pressed without the field. The compacts are dried and fired at 1250°C. During sintering, densification must occur without significant grain growth.

Because the cost of hard ferrites is much lower than metallic permanent magnet materials, they have found wide use in large-volume, low-cost consumer electric applications. These include arc segments for small electric motors for automobiles and portable electric appliances and for loud speakers. These uses account for the greater part of the volume. Other uses are for ore separation, microwave applications, magnetic recording applications, and as plastic-bonded material for sealing refrigerator doors.

Soft Ferrites

by John B. Ings, Ceramic Magnetics, Inc.

Metal magnets are limited in use to low-frequency applications by virtue of their low electrical resistivity. As the alternating-current frequency increases, eddy currents (circular currents in a material that oppose those created by the primary current) start to flow through the magnets. Eddy currents are a large source of magnetic losses at high frequencies. In the 1930s work in the Netherlands by Snoek and co-workers resulted in a new class of materials called ferrites, typified by $MnZnFe_2O_4$. These materials exhibit both high permeability and electrical resistivity, which allows magnets to be used in high-frequency devices.

This class of ferrites is widely used today in high-frequency electronic applications. Applications such as miniaturized power supplies for computers, isolation transformers to protect computer networks from stray voltage spikes, and magnetic read/write heads for computer back-up tape drives are driving the market with literally a billion parts per year being produced worldwide. The largest growing segment of this market is high-frequency switched-mode power supplies. According to the Faraday induction equation for an alternating current, the voltage on the secondary winding of a transformer core is

$$E = 4.44B_m^2 NAf \times 10^{-8}$$

where E is the voltage (volts), B_m the maximum induction (gauss), A the cross-sectional area (cm^2), and f the frequency (hertz). This shows that, if the frequency is increased from 60 to 60,000 Hz, it is possible to reduce the size of the core by a factor of 1000. This would allow the replacement of the old, bulky power transformers seen in 1960 vintage TVs with a part that can fit onto a printed circuit board. However, the low electrical resistance of metal magnets has hindered this switch to high-frequency operation. As the frequency of operation increases, eddy currents appear. The eddy current losses are proportional to the square of the frequency:

$$P_e = KB_m^2 \, f^2 \, d^2/\rho$$

where P_e is the eddy current loss (W·cm^{-3}), K a sample shape constant, d the shortest dimension perpendicular to the flux path (centimeters), and ρ the resistivity (Ω·cm).

Although metal magnets have an excellent saturation magnetization, they are conductors of electricity. This conductivity causes tremendous losses at high frequency that cause the cores to heat up. In contrast, the electrical resistivity of $MnZnFe_2O_4$ is about 10^6 times less than the best metal magnetic materials. It is the unique magnetic and electrical properties of $MnZnFe_2O_4$ that allows for the existence of miniaturized power supplies in TVs, computers, and consumer electronics. Without this material, the recent advances in computers and consumer electronics would not have occurred.

Magnetic Recording Media

by Robert J. Youngquist

There are many products that are composites containing ceramics as their key material. One of these is magnetic recording tape that is used to record sound, visual images, and encoded information, such as for computer memories. Poulsen invented magnetic recording at the turn of the century, and in WWII the Germans developed the "magnetophone," the first practical recorder. In 1944 Brush Company asked 3M Company to develop a magnetic recording tape. Since the time of its commercialization by 3M in the 1940s, these composite products have been made in many other forms, such as disks or diskettes. Substantial improvements have been made in the information density that can be recorded and retrieved by controlling the particle size and shape as well as the stoichiometry of iron oxides and additives to the compositions. One of the early improvements was the use of acicular γ-iron oxide as the recording medium, coated on polyester film. Later, chromium dioxide and many other "alloys" of oxides were made to improve the medium or provide special recording capabilities. Careful coating technology was developed to provide error-free uniformity and minimize wear on the recording heads. These products represent a good example of the careful interaction between the device makers and the tape producers so as to optimize the quality of the sound, data, or images that were the end result. This was particularly true as recording heads were developed along the way that could provide even better resolution. In 1956 Ampex introduced the data cartridge system that is the preferred system for computer data backup, and a year later the Betamax and VHS systems were introduced. As these and other new systems came along, such as 32-channel audio-mastering recording equipment, the demands of the ceramic component of the composites became ever more stringent.

Composites such as magnetic recording media are examples of a major trend in advancing ceramic technology where the active ingredient is a ceramic. Other examples include electronic substrates, optical reflecting films (such as are used in highway signs), dental fillers for tooth repair that provide not just color and mechanical properties but subtle surface reflections, and plastic magnets for refrigerator art.

Crystal-Oriented Hot-Pressed Manganese Zinc Ferrites

by Koichi Kugimiya, Matsushita Electric Company

Hard disk drives and video cassette recorders are now very popular. Over 150 million and 50 million, respectively, are now produced annually worldwide. Initially, in the 1960s, the magnetic heads of these devices were generally composed of magnetic metals, but these were gradually replaced by ferrites, because the latter are harder and more durable in service. As higher recording densities were developed, higher head qualities were required, including lower porosity and higher magnetic flux. One of the best solutions was to use a hot-pressing technology of manganese zinc ferrites instead of the conventionally sintered nickel zinc ferrites.

A special hot press with unique features is used. A special silicon carbide mold and press bar were developed that could withstand loads of 500 kg/cm^2 at 1400°C. To prevent reaction between ferrite materials and molds, alumina powder is packed fairly tightly between them. Thus, the specimen is subjected to pseudoisostatic pressure, although a uniaxial pressure is applied. Because of the slow air flow through the packed alumina and the fairly strong reducing atmosphere of the silicon carbide molds at high temperatures, it was possible to sinter ferrites with no modification of the furnace atmosphere as is generally required for sintering such ferrites. Densification and grain growth are completely and independently controlled, giving additional freedom to control the microstructure of the sintered material.

In 1976, the VHS–VCR was introduced; this was the first long-play VCR for home use. The new model required even higher recording density by almost 2 times compared to the previous model and even better durability for assurance of longer service for the general home users. A crystal-oriented ferrite was used for the first VHS model. Because its unique anisotropy of mechanical and magnetic properties, one could take advantage of the best combination of anisotropic properties to satisfy the difficult head requirements. Magnetic heads today are composed of complex layered structure materials: magnetic metals, ferrites, and other ceramics.

The manufacturing process of a crystal-oriented ferrite is unique. Powders of thin, flat shapes are mixed without fracture, shaped by a wet-molding method that squeezes out water and aligns particles, and hot-press reacted and sintered to form spinel bodies in which (111) crystallographic axes of grains are aligned parallel to the pressing direction. The alignment is due to the nonequiaxed shape of the powder particles, hematite and hydroxo-manganese oxide, and the topotactic reactions between them.

Optical Ceramics and Glasses

AlON Transparent Ceramic (Aluminum Oxynitride)

by Edward A. Maguire, Raytheon Electronic Systems

AlON (aluminum oxynitride) is an extremely durable, transparent ceramic material conceived at the U.S. Army Research Laboratory and developed at Raytheon Company. It has excellent optical transparency in the near ultraviolet, visible, and infrared regions of the spectrum up to approximately 5 µm in wavelength. The material features mechanical and optical properties similar to sapphire with the advantages of an isotropic cubic crystal structure. It has the nominal composition $Al_{23}O_{27}N_5$ and is polycrystalline.

AlON is manufactured by conventional powder processing techniques. It can be fabricated in plate, dome, rod, and tube shapes in a wide range of sizes by a variety of methods, including injection molding, dry pressing, isostatic pressing, extrusion, and slip casting. The material is easily polished to fine surface finishes by normal cutting, grinding, and polishing operations.

Figure 19. *Stages in fabrication of transparent AlON. Courtesy of Raytheon Co.*

AlON finds application as optical windows and domes, sight tubes, etc., where transparency and high durability are important. Its resistance to abrasion, rain erosion, and solid-particle impacts is superior. It is a premier transparent armor material with potential use in fixed structures, land vehicles, aircraft, and ships as well as in individual personnel protection, such as face shields, windows, and view ports.

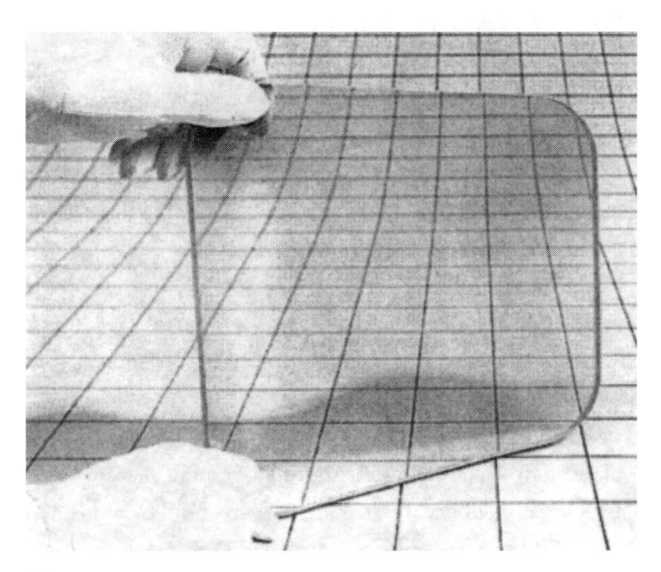

Figure 20. *AlON transparent armor window. Courtesy of Raytheon Co.*

Chemically Vapor Deposited ZnS and ZnSe

by Bernard A. diBenedetto, Raytheon Electronic Systems

The rapid growth of the commercial and military infrared optical industry during the 1960s and 1970s created a need for new lens and window materials. Optical-quality ceramics for these applications need to be free of pores and impurities to minimize scatter and absorption. Conventional ceramic processing had to be supplanted by single-crystal growth, hot pressing, and other processes to achieve the required properties. In order to meet this growing need for large windows, Raytheon Research Division expanded its materials capability to develop a family of infrared-transmitting materials.

The process chosen for the preparation of the primary material compositions of interest, ZnS and ZnSe, was the chemical vapor deposition process in order to avoid sublimation of constituents that occurred during normal sintering processes. The chemical vapor deposition process consisted of reacting zinc vapor with H_2S or H_2Se gas to form solid deposits of ZnS or ZnSe . The deposit built up over time, typically yielding several plates 36 in. × 48 in. (up to) 1.0 in. thick. The configuration of the deposited product replicated the shape of the chamber walls; therefore, complex-shaped parts, such as domes, cylinders, rods, and lenses were fabricated for a variety of applications where the size and shape were key. Production size deposition furnaces were specially designed and built complete with computer control in order to meet the growing need for these materials. Although the process was nearly 100% efficient, much of the processing hardware had to do with environmental issues concerning potential exhausting of toxic gasses.

The extensive defense industry use of Raytheon's chemical vapor deposited ZnS and ZnSe created a new, $20 million/year infrared optical materials industry. The availability of these materials at reasonable cost spawned their use in commercial, police, and industrial security systems, leading to growth of these industries. One of the most successful examples is the extensive use of ZnSe lens components in robotics industrial lasers for welding in the automotive industry.

Further improvement was realized in ZnS material by a postdeposition high-temperature and high-pressure treatment (HIP) that was developed to reduce the visible and midwave optical scatter. More recently

Figure 21. *Transparent dome. Courtesy of Raytheon Co.*

improvements were made to the chemical vapor deposition process itself, to serve additional applications in the infrared midwave range.

Uncooled Infrared Cameras

by Bernard M. Kulwicki, Texas Instruments, Inc.

Field-biased, pyroelectric, barium strontium titanate (BST) ceramic detector material forms the basis for hybrid focal plane arrays (FPAs) that operate at room temperature and exhibit an image resolution of 320 × 240 pixels and noise equivalent temperature difference (NETD) as low as 0.04 K. The technology for producing the detector arrays and integrating them into thermal imaging systems was developed and commercialized by Texas Instruments Defense Systems Group over the past 20 years. This business was acquired by Raytheon Corporation in 1997 and is now part of Raytheon TI Systems (RTIS) in Dallas, TX. Infrared cameras using this technology were introduced on the market in 1996. Both defense and commercial night vision applications are being served.

Manufacture of each detector requires the production of an array of 80,000 BST pixels, which are individually 50 μm × 50 μm in area by

20 μm thick. The FPA is made by bump-bonding the BST pixels to a corresponding array of circuit elements on the readout integrated circuit (ROIC), using polyimide mesas for thermal isolation.

Production of the ceramic pixel array begins with a 150 mm diameter × 0.7 mm thick (SEMI format) BST "wafer," 99% dense, with a grain size of 1.5 μm, polished on one side. Forty-two pixel arrays are created on the top surface by laser reticulation. After slag removal and reoxidation, the kerf regions are backfilled with chemical vapor deposited (CVD) polymer, the surface is replanarized, and the common electrode and optical coating are applied.

The individual dies are cut out, thinned, and polished from the back side, and the pad metal and bond metal (indium) are applied. After the CVD polymer is removed by reactive-ion etching, the BST array is ready for attachment to the ROIC. Bump-bonding is currently achieved at high yield with typically >99.9% good pixels.

The specific technology that enables this fabrication complexity, while at the same time retaining excellent electrical performance, involves donor doping the $(Ba,Sr)TiO_3$ with 0.5–1.0 mol% of a trivalent ion on the barium sublattice (U.S. Pat. No. 5,314,651). With proper control of composition, purity, and homogeneity as well as careful processing, it has become possible to achieve high ceramic density and fine grain size together with outstanding dielectric and pyroelectric performance: direct current resistivities exceed 10^{13} Ω·cm and loss-limited pyroelectric figure-of-merit exceeds 0.13 $(cm^3/J)^{1/2}$ under bias.

Development of the detector ceramic to withstand the rigors of thinning, reticulation, and bonding while maintaining satisfactory electrical performance and development of the processes for conversion of the ceramic to the pixel array, forming the hybrid structure with the ROIC (together with design of the ROIC, and integration of the hybrid detector with the system electronics and the optics) represented formidable technical challenges. Today practical, low-cost, uncooled infrared cameras are manufactured routinely.

Electrooptic Ceramics

by Gene Haertling, Clemson University, emeritus

Ceramics have long been known for their desirable artistic, structural, thermal, electrical, and dielectric properties; however, it has only been

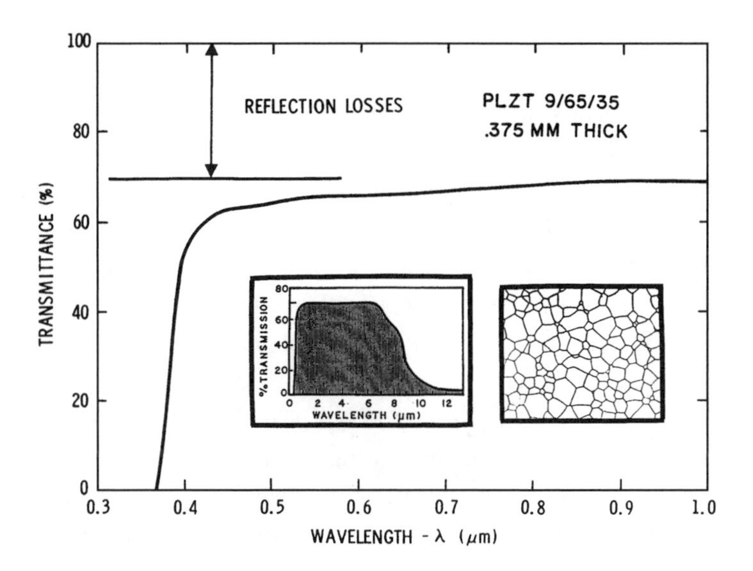

Typical Optical Transmission Curves and Microstructure of PLZT Electrooptic Ceramics

Figure 22. *Typical optical transmission curves and microstructure of PLZT electrooptic ceramics. Courtesy of Sandia National Laboratories.*

since 1967 that they have proved to be viable candidates for applications in the high-technology fields of optics and electrooptics. Competing in a world where single crystals have reigned exclusively for many years, electrooptic ceramics have grown from a totally new concept in optical processing and light modulation to where it is now a relatively mature technology that is recognized and utilized in numerous military and commercial products sold throughout the world.

Among the several factors that were instrumental in achieving these results, at least three can be singled out for special mention. These are (1) clarity, (2) composition, and (3) characteristics. Regarding clarity, electrooptic ceramics are fully dense, optically transparent materials (0.4–6.5 μm) that are fabricated from chemically pure, coprecipitated, oxide powders via either oxygen hot pressing or special sintering techniques. The pore-free, optical quality of these materials is a must for their successful utilization and application.

Regarding composition, they consist of special formulations not found in nature, but rather, are synthetically compounded to produce highly responsive materials with optimal electrooptic coefficients and possessing a rather large array of unique electrooptic effects involving optical retardation, light scattering, and photoferroelectric phenomena. Most notable of all of the electrooptic ceramic compositional systems developed over the years by various investigators is the PLZT (lead lanthanum zirconate titanate) solid-solution system. The PLZTs are well-known for their ease of manufacture, compositional flexibility, excellent transparency, variety of electrooptic effects, high electrooptic coefficients, and operational reliability. Because they are unique in many respects, they have remained the standard of the industry since their development by Haertling in 1969. A typical optical transmission curve, microstructure, and samples are illustrated in the accompanying figure.

In regard to characteristics, electrooptic ceramics possess these properties because they belong to a unique class of materials known as ferroelectrics. As such, they possess a spontaneous internal polarization (dipole moment) that derives from its unit cell containing atoms (ions) at specific locations within the cell; these atoms, as well as their orbiting electrons, are readily influenced by the application of an electric field via suitable electrodes. As polarized light interacts with these atoms and electrons, it is split in two orthogonal waves whose directional velocities are changed in response to the electric field, thus yielding an optical delay or retardation between the two waves, which, when resolved by a second polarizer, results in a shuttering action (as in a light valve) when the voltage is turned on or off (digital operation). Analog (nondigital) operation also can be realized, because the total amount of retardation is proportionally dependent on the strength of the electric field. This effect is used to produce black, white, shades of gray, or any of the spectral colors. Other useful characteristics accompanying ferroelectricity in these materials are piezoelectricity (stress-induced voltage or voltage-induced strain) and pyroelectricity (temperature-dependent polarization).

Applications for the electrooptic PLZT ceramic technology were first developed by Sandia National Laboratories in the mid-1970s with commercial devices following a few years later. These include (1) the U.S. military thermal/flash eye protective devices for the SAC B-52, FB 111, 747 Flying Command Post, B-1B bomber, KC-135 and (2) commercial devices, such as the Bell and Howell optical data recorder, Kodak optical processing and image enhancement systems, Light Valve Technologies image recording systems, Minolta copiers, and Steiger SA (Swiss) offset printers.

Glass Lasers

Glass and Ceramic Lasers

by Matthew J. Dejneka, Corning Incorporated

Schawlow and Townes first suggested the idea of an optical laser in 1958. Two years later, Maiman demonstrated the first experimental laser using a single crystal of ruby (Cr^{3+}-doped corundum), and several single crystal lasers, as well as the helium–neon gas laser, were reported the following year.

Activator-doped single crystals generally have greater cross sections and narrower fluorescent linewidths than similarly doped glasses. Glass, however, can be formed in large meter-sized pieces as well as tiny micrometer-sized fibers, and is much more economical to produce. Glass lasers also are well suited for high-power, short-pulse lasers and amplifiers because of their lower cross sections.

The first glass laser was made by Snitzer in late 1961. He used a Nd^{3+}-doped barium crown glass rod with a refractive index of 1.54, clad with a soda–lime–silica glass with a refractive index of 1.52, and pumped with a xenon flashlamp. The cladding was used to form a light guide and improve the mode properties to compensate for the poor optical quality of the glass rods obtained from the small 1 lb melts used. Thus, not only was it the first glass laser, but it also was the first waveguide laser as well, with a background loss of 0.15 dB/cm. Snitzer then went on to demonstrate the first fiber amplifier using the same glass pulled to a 1.5 mm fiber with 25 μm core. Using a 370 J pump pulse he was able to obtain a net signal gain of 5.3×10^4 (47 dB) in a 1.0 m long Nd^{3+}-doped fiber amplifier, a decade before the invention of low-loss optical fiber. Stimulated emission was then demonstrated in Yb^{3+}-, Ho^{3+}-, Yb-Er^{3+}-, Tm^{3+}-, and Tb^{3+}-doped glasses with most of the recent work shifting back toward Nd^{3+} for high-power lasers for fusion. The Nd^{3+} glass laser is the only commercially important glass laser, because its high-efficiency, room-temperature operation.

The first polycrystalline laser was demonstrated by Carnall *et al.* in hot-pressed CaF_2 doped with Dy^{3+}. Anderson then patented the cold-pressed and sintered Nd:YAG (yttrium aluminum garnet) ceramic laser. The main problem with early ceramic lasers was their high loss and threshold pump power due to scattering, which made them inferior to their single-crystal and glass counterparts. Anderson's ceramic laser then was improved upon by Greskovich and Chernoch in the 1970s and then recently perfected by Ikesue and co-workers, who achieved single-crystal performance in a transparent YAG ceramic laser with a slope efficiency of 28%. Ikesue *et al.* synthesized high-purity starting materials (>99.99%) with particle sizes < 2 μm, mixed them with an ethyl silicate sintering aid, isostatically pressed 16 mm disks at 140 MPa, and vacuum sintered them to transparency at 1750°C. These ceramic lasers offer more economical processing than single-crystal materials and a thermal conductivity > 5 times greater than glass.

Transparent glass-ceramics can provide the best (or worst) of both glass and crystals. If the activator ions are incorporated into the crystals, the high cross sections and narrow linewidths of the crystal can be achieved with the formability and processing ease of a glass. The other option is to design a system in which the activator is retained in the glassy phase to take advantage of the increased toughness and thermal conductivity of some glass-ceramics. Anything in between these two extremes also is possible. The first glass-ceramic lasers were demonstrated independently in early 1972 by Muller and Neuroth of Schott, and by Rapp and Chrysochoos of the University of Toledo with both groups reporting lasing in transparent β-quartz glass ceramics. Unfortunately, the glass-ceramics had thresholds 2–3 times as high as the untreated glass and efficiencies an order of magnitude lower. This was due to the fact that rare-earth ions have large ionic radii with a strong preference for eightfold and higher coordination and, hence, poor solubility in β-quartz, which contains only small tetrahedrally coordinated cations. Thus, the Nd^{3+} ions concentrated in the remnant glassy phase, resulting in concentration quenching and poor performance.

Luminescence and even upconversion from active ions has been reported in many transparent glass-ceramics in which the active ions are well partitioned into the crystalline phase, unlike the β-quartz glass-ceramics. Lasing has not yet been achieved in any of the materials mentioned below, but they are worth mentioning to show the rich possibilities for future glass-ceramic lasers. Andrews, Beall, and Lempicki showed that the luminescence quantum yield of Cr^{3+} in mullite glass-ceramics increased by 2 orders of magnitude when the glass is cerammed and that Cr^{3+} partitioning into the mullite crystals is 99%. Wang and Ohwaki demonstrated the first transparent oxyfluoride glass-ceramics and Er^{3+} upconversion from 980 to 545 nm that increases 100-fold upon ceramming these novel materials. Lastly, new transparent LaF_3 glass ceramics have been developed that show a twofold increase in 1.3 mm Pr^{3+} fluorescence efficiency over fluoride glasses and simultaneous red, green, and blue emission from Eu^{3+}.

Glass Lasers

by Emil W. Deeg, Lemoyne, PA, AMP, Incorporated, retired

Prior to and for some time after a 1961 publication by E. Snitzer of the American Optical Company appeared that claimed to have found laser action in a neodymium-doped barium crown glass, it was thought impossible that glass could serve as laser material. The reason for this opinion was a combination of adhering to the outdated and over-simplified random network theory of the glass structure and of unfamiliarity with the fact that population inversion can take place in certain ions at energy levels representing inner, nonvalence shells. It was generally accepted that the random glass network, being responsible for broadening of absorption and fluorescence lines of ionically colored glasses, would prohibit the formation of pump and emission levels sharp enough for stimulated emission.

Publication of Snitzer's innovation initiated worldwide efforts to develop laser glass compositions. Efforts concentrated on neodymium-doped glasses emitting laser radiation at the near-infrared wavelength of 1.06 μm. Such glasses were to be free of striae and had to have very small local variations of the refractive index. Optical glass manufacturers selected originally base glass compositions that were already produced on a large scale in high quality and at relatively low cost. Examples are ophthalmic crown glasses and the widely used BK7. Other experimental base glasses included alkali (alkaline-earth) silicates, alkali (alkaline-earth) aluminosilicates, borosilicates, aluminoborosilicates, lithia–alumina–phosphates, (rare-earth) silicates, and (rare-earth) borates. Rare-earth elements other than neodymium potentially useful as dopants in laser glasses were studied at various research laboratories. Laser action was reported for glasses containing erbium, gadolinium, holmium, and ytterbium. Most of them had to be operated at temperatures below 100 K. Laser action also was found in glasses doped with two rare-earth elements (Yb/Nd, Er/Yb, Nd/Ce) or the combination Nd/U.

Today, about 35 years after demonstrating feasibility of glass lasers at several institutions, only neodymium-doped and, to some extent, erbium-doped glasses are important. Erbium-lasers are less efficient than neodymium-lasers but are of interest because of their emission wavelengths at 1.54 and 2.9 μm. The first wavelength is important because it allows laser amplifiers in form of fibers for long-distance fiber-optic communication. The second wavelength falls in the H_2O absorption bands and makes erbium-lasers, particularly as erbium–yttrium aluminum garnet, useful for medical applications. Neodymium-doped glasses are most prominently used in amplifiers for high-power, pulsed-laser systems. The most popular base glass today is a lithia–alumina–phosphate glass. It is chemically durable, provides desirable laser properties for neodymium and can be toughened efficiently by ion exchange.

Optical systems of specific designs require materials with properties optimized for the specific application and optimization of the design itself. The 35 years of glass laser development offer many examples. The initial glass lasers consisted of a neodymium-doped glass rod similar to that shown in Fig. 23.

Figure 23. *Small laser rod. Schott LG55 base glass. Courtesy of University of Rochester.*

Both ends of such a laser rod were ground and polished parallel to each other within very tight tolerances. Subsequently, multilayer dielectric mirrors were applied to these end faces, so that each rod became its own Fabry–Perot etalon. There also were setups where the mirrors were separated from the rod, an arrangement most common today. The rod was pumped by either a helical flash lamp with the rod positioned at the axis of the helix or a linear flash lamp mounted parallel to it. Already in the early development phases, rod and linear flash lamp were placed at the focal lines of cylindrical mirrors with elliptical cross section. This setup continues to be used in small lasers used for machining. Particularly in high-power lasers, the rods in Fabry–Perot cavities were replaced by laser glass slabs. In such arrangements the beam is kept inside the laser glass by total internal reflection. Arrays of linear flash lamps pump the disk through its top or/and bottom surface. The pump light for neodymium-doped glasses or YAG usually covers a range from ~ 300 to 900 nm. Broadband xenon-flash lamps are commonly used as pumping sources for pulsed mode operation, krypton-flash lamps in the few cases where cw-mode operation of neodymium-doped glass is considered.

The OMEGA laser system of the Laboratory for Laser Energetics of the University of Rochester serves to illustrate the present state of glass lasers. Instead of the modest laser rods represented by Fig. 23, noticeably larger rods now are used. The OMEGA laser system includes four stages of rod amplifiers and two stages of disk amplifiers. Before it reaches the amplifier, the initial laser beam is split three ways. Each of these beams is then amplified and split in five, resulting in 15 beams that are again amplified. The first two stages of the amplifier use 64 mm diameter rods, the third 90 mm diameter rods. Each of the 15 beams then is split four ways, resulting in a total of 60 beams, each of them passing through a second 90 mm rod amplifier, before being amplified in a 150 and 200 mm disk amplifier. These beams are frequency tripled by passing them through two KH_2PO_4 crystals (KDP) in series, resulting in high-power ultraviolet radiation at 351 nm. The OMEGA system has produced >40 kJ in the ultraviolet and irradiated targets with up to 37 kJ. It allows generation of intensities of 10^{17} W/cm^2, which is sufficient to transform a solid target to plasma of 10^7–10^8 K. Thirty-five years ago we were elated to have a glass laser beam produce a tiny black spot on a sheet of paper.

In the early 1960s the laser heads were about 6 in. (15 cm) long and 4 in. (10 cm) in diameter. The facility, which includes the area where the target to be exposed to the laser beams is mounted, is about 27 m wide, 66 m long, and 4.9 m high. The early laser heads were held by laboratory clamps or mounted in a microscope stage. The floor supporting the laser serves as an optical table, rests on a bed of gravel, and is structurally independent of the building that houses the facility.

Low-Loss Optical Fiber

by John B. MacChesney, Bell Laboratories, Lucent Technologies

Since the inception of the telephone, engineers have sought means to increase bandwidth (capacity), which is proportional to the signal's frequency. Light provides means to signal at frequencies to terahertz (10^{12} Hz). By the 1960s, technology was in place: semiconductor laser light sources, pulse-code modulation, and understanding of how light propagates in thin glass fibers consisting of a central core whose refraction index is slightly higher (~1%) than that of the surrounding cladding. All that remained was to find a glass of sufficiently low attenuation.

Extensive research ensued toward purification of glass-forming constituents of common glasses (pyrex, soda–lime–silica) and yielded fibers that were satisfactory for the most primitive communication systems. However, they were never able to achieve part-per-billion levels of transition-metal oxides or OH^- below or even near 1 ppm needed for optimum transparency. More fundamentally, the cation–oxygen vibrational absorptions limited the spectral range for transmission to 0.9–1.0 μm. The best loss was on the order of 5 dB/km.

The discovery of means to fabricate high-silica glass fiber changed all that. First, vapor deposition reduced transition-metal contamination to the ppb range, the spectral range was extended to 1.6 μm where scattering loss via Rayleigh scattering is decreased by ~ 5 and Si–O vibrational modes did not interfere. In the present stage of development, OH^- is reduced to the tens of ppb, where the broad absorption peak at 1.38 μm does not contribute to loss at the 1.3 or 1.5–1.6 μm transmission range. Commercial fiber today has a loss of 0.33–0.35 dB/km at 1.3 μm and 0.19–0.2 dB/km at 1.55 μm. This is sufficient for unrepeated transmission spans of >100 km.

These fibers consist for the most part of vitreous silica with a core-index increased by germanium doping. Processing took two routes: inside—modified chemical vapor deposition (MCVD) and plasma vapor deposition (PVD); outside—outside vapor deposition (OVD) and vertical axial deposition (VAD). MCVD, invented at Bell Laboratories in 1974, operates with a silica tube mounted in a glass-working lathe. Reactants (SiC_{14}, GeC_{14}, O_2) flow into the tube and are heated by an oxyhydrogen torch traversing along the tube on the outside. Colloidal-sized particles deposit on the wall downstream of the torch and consolidate to a vitreous silica layer as the torch moves past them. After many layers are deposited, the reactants are stopped and the tube collapsed with the deposited layers forming core and cladding. This rod of high-silica glass is drawn to fiber at ~2100°C. PCVE, developed by Philips, is similar except that the vitreous layer deposited on the inside of the tube is formed by nonisothermal plasma created in an atmosphere of reactants at a few torr (few hundred pascals) pressure by a microwave cavity (traversing along the tube) operating at 2.45 GHz.

Outside processes react chloride reactants in a torch fueled by hydrogen or propane to create submicrometer-sized particles. Corning announced OVD in 1973. Here, particles are deposited on a mandrel-doped core first, followed by cladding. This forms a cylindrical boule that is removed from the mandrel, consolidated in a chlorine-containing atmosphere to remove OH^- created in the flame, and subsequently drawn to fiber. The other outside process, VAD, announced by Nippon Telephone and Telegraph in 1977, uses two torches to deposit end-on. One torch deposits core on a rotating mandrel, which is slowly withdrawn from the vicinity of the torch. Pure silica is deposited on the porous, partly sintered boule as it passes a second torch positioned above the first. When complete, the porous body is consolidated to produce a vitreous preform that is then drawn to fiber.

Modern processing produces preforms of up to 1000 km length of fiber (30 kg). These are created, in the case of the inside processes, by over-cladding the core-rod with a vitreous silica cylinder produced by vapor deposition or more recently by sol–gel. Outside-deposited core-rods are overclad by rapid flame deposition of silica.

These processes—worldwide—have combined in 1997 to produce ~40×10^6 km of fiber comprising 1200 metric tons of silica.

Erbium-Doped Optical Fiber Amplifiers

by Kenneth Walker, Lucent Technologies

Erbium-doped fiber amplifiers (EDFA) have revolutionized optical communications. Optical-fiber communication systems have been in use since the early 1980s. The advantage of optical fibers is that they are theoretically capable of supporting extremely high bandwidths (many teraherz). One serious limitation of optical fibers is that, despite the extremely high purity and transparency, the light attenuates with distance, limiting practical lengths to ~100 km. At this point an optoelectronic regenerator is necessary to detect the light, perform high-speed electronic processing, and generate a new optical signal to send through the next span of optical fiber. These regenerators represented a serious bottleneck, limiting the practical speed of systems. The invention of the erbium-doped fiber amplifier at both the University of Southampton and Bell Laboratories in 1987 has dramatically transformed optical communication systems.

These amplifiers consist of a short length (~20 m) of optical fiber whose core is doped with erbium (500 ppm). Constant power pump light at either 980 or 1480 nm is combined with the modulated signal light and injected into the doped fiber. The pump light raises the erbium ions to an excited state. The optical signals (at wavelengths between 1530 and 1610 nm) induce stimulated emission in the erbium ions and are thereby amplified. Erbium-doped amplifiers typically amplify the optical signal by a factor of 100 to 10,000. A key feature of the amplifiers is that they work at any bit rate, and they can simultaneously amplify many optical signals at different wavelengths (1530–1610 nm). Optical-fiber amplifiers are used extensively in long-distance communication systems. Some of the commercial systems use as many as 40 wavelengths of light, with each wavelength transmitting at 10 gigabit/s. This compares to typical systems in the early 1990s using only a single wavelength at 2.5 gigabit/s.

Ultraviolet-Induced Refractive Index Changes in Glasses

by Thomas A. Strasser, Bell Laboratories

In the future, structures that guide light will be capable of performing complex functions that are comparable analogues to electrical-integrated circuits. These devices will perform critical tasks in communications systems and probably some in computing systems and information storage technologies as well. One crucial capability that will be needed is a means to control the refractive index of materials on the scale of a wavelength of light (about 1/100th the diameter of a hair). A technology that is currently being commercialized demonstrates this capability within a glass fiber by creating atomic defects in the fiber with an ultraviolet-(UV-) radiation-induced chemical reaction. These defects change the density of the glass as well as the absorption properties, both of which contribute to refractive index changes that can be as large as 0.01 in silica fibers.

Hill was the first to discover this process in fibers by imprinting a phase grating in 1978, where his group reported reduced transmission of an argon-ion laser travelling through a fiber. The cause of this reduced transmission was a weak, periodic variation in the refractive index along the fiber that formed a reflection diffraction grating. This grating was the result of a periodic photochemical reaction that was initiated by a standing wave pattern of the laser interfering with its own reflection from the exit facet of the fiber. In the late 1980s Meltz and Morey showed that these photochemical reactions were more efficiently driven with more flexible control of the periodicity using UV radiation from the side of the fiber at twice the energy of that used in the original gratings. Finally, in 1993 Atkins and Lemaire reported the ability to increase achievable index changes by up to 100 times or more by loading the structure with molecular hydrogen before UV exposure.

The first commercial realization of this technology was to create UV-induced phase gratings in optical fibers for sensor and telecommunication systems. These gratings are unique in their ability to reproducibly create selective reflection or loss at wavelengths that are resonant with the local periodicity. Fiber uniformity, as well as precise control of the periodicity and exposure of the UV light, has enabled this technology to fabricate the most discriminating, low-loss optical filters available by any technology known at this time. In the future, the commercial impact of this technology will extend to new devices, such as direct writing of two-dimensional waveguides on silicon wafers as was originally demonstrated by Mizrahi. In addition, new glass material systems will be used beyond the conventional germanium-doped silica. In this area, photosensitivity already has been demonstrated in phosphorus-doped silica and chalcogenide glasses. It is clear from the achievements to date that many innovations impacting new applications will result from this ability to photochemically engineer local refractive index and will provide higher functionality in future generations of optical devices.

Sensors

Positive Temperature Coefficient Resistors

by Bernard M. Kulwicki, Texas Instruments

Positive temperature coefficient (PTC) resistors are prepared from donor-doped, ceramic $BaTiO_3$ and the solid solutions $(Ba,Sr)TiO_3$ and $(Ba,Pb)TiO_3$. The ceramics become semiconducting in a narrow range of doping concentration, typically 0.2–0.5 mol% of either a trivalent element (e.g., lanthanum or yttrium) residing on the barium sublattice or a pentavalent element (e.g., niobium) residing on the titanium sublattice. When sintered in an oxidizing atmosphere, they have baseline resistivities between 10 and 10^4 Ω·cm. At temperatures above the ferroelectric–paraelectric transition temperature, or Curie point (T_c), the resistivity increases by a factor of 10^3–10^5. The Curie point is ~120°C for $BaTiO_3$. Substitution of strontium on the barium site reduces the Curie point such that $(Ba_{0.6}Sr_{0.4})TiO_3$ transforms around 0°C; substitution of lead on the barium site increases the Curie point such that $(Ba_{0.6}Pb_{0.4})TiO_3$ transforms at ~300°C. Acceptor counterdoping (e.g., 0.04–0.10 mol% manganese) enhances the resistivity increase above the Curve point, and liquid-phase-forming additives (e.g., TiO_2, SiO_2) permit control of grain growth and electrical performance.

Sophisticated manufacturing techniques have evolved for simultaneously controlling the electrical and mechanical properties of PTC ceramics. They involve precisely controlling microchemistry and microstructure, minimizing undesirable impurities, and controlling the sintering process with special attention to oxidation and electrical uniformity of the sintered product. These controls enable applications that require the PTC resistor to withstand high electric power insertion over lifetimes that may exceed 10 years and 10^6 electrical cycles.

The resistivity behavior of PTC ceramics has been found to occur at the grain boundaries of the ceramic, where electrical junctions, akin to back-to-back Schottky barriers, exhibit a strong temperature dependence of the barrier potential, through its interaction with the nonlinear polarization and dielectric properties of the matrix. Hence, the resistivity exhibits both a voltage and frequency dependence that are functions of the grain size. These can be application design constraints.

The PTC phenomenon was discovered by P. W. Haayman and coworkers at Philips Research (German Pat. No. 929,350, 1955), and major contributions to understanding the properties and the physical mechanism were made by G. H. Jonker (Philips), W. Heywang (Siemens), and O. Saburi (Murata) in the late 1950s and early 1960s. The first products, introduced on the market around 1965 were an oil level sensor (Siemens) and a diode oven (Texas Instruments).

Applications of PTC ceramics include temperature sensors, e.g., for electric motor protection; liquid level and air flow; self-regulating heaters in automobiles and appliances; and current-switching devices in refrigerators and air-conditioning systems. Major manufacturers of devices utilizing this technology also include Murata, Matsushita, TDK, Siemens, and Philips.

Smart Electroceramics

by Robert E. Newnham, The Pennsylvania State University

Smart materials function as both sensors and actuators. This combination of sensing and actuating mimics two of the functions of a living system—namely, being aware of the surroundings and being able to respond to that signal with a useful response, often in the form of a motion. Smart materials sometimes have a control system and sometimes not.

Some are passively smart and function similar to the reflex responses of the human body. Ceramic varistors and positive temperature coefficient (PTC) thermistors are passively smart materials. When struck by high-voltage lightning, a zinc oxide varistor loses most of its electrical resistance, and the current is bypassed to ground. The resistance change is reversible and acts as a standby protection phenomenon.

Barium titanate PTC thermistors show a large increase in electrical resistance at the ferroelectric phase transition near 130°C. The jump in resistance enables the thermistor to arrest current surges, again acting as a protection element. The voltage-dependent behavior of a varistor and the temperature-dependent behavior of the PTC thermistor are highly nonlinear effects that act as standby protection elements that make the ceramics smart in a passive mode.

Actively smart ceramics utilize a control system and are analogous to the cognitive responses in a living system. Here the sensed signal is analyzed—perhaps for its frequency components—and the appropriate response is made through a feedback system. Examples of actively smart materials include vibration-damping platforms for outer-space telescopes and electrically controlled suspension systems for automobiles. In applications such as these, the direct piezoelectric effect (stress-induced polarization) is used in sensing vibrations, and the converse piezoelectric effect (field-induced strain) creates the actuator response.

Four major families of actively smart materials have been developed during the past few years: piezoelectric and electrostrictive ceramics typified by $Pb(Zr,Ti)O_3$, and $Pb(Mg,Nb,Ti)O_3$, usually referred to as PZT and PMN, and magnetostrictive and shape-memory alloys, such as terfenol, $(Tb,Dy)Fe_2$, and nitinol, NiTi. All four materials are primary ferroics (ferroelectric, ferromagnetic, or ferroelastic) with domain-wall motions that assist in the sensing and actuating functions. Two phase transformations are involved in most of these smart materials. One type has a high Curie temperature as in PZT or terfenol, and the device is operated near a second transition where the electric or magnetic dipoles are about to undergo a change in orientation. The second type involves a partially ordered phase, as in electrostrictive OMN or the shape-memory alloys. These materials operate near a diffuse phase transition between two coexisting phases, the high-temperature or austenite-like phase and a low-temperature or martensite-like phase.

Another approach to making smart materials is to bring together two or more different materials, each of which has a phase transition or instability associated with it. In our transducer program, we combine polymeric materials having phase transitions accompanied by large changes in elastic properties and ferroelectric ceramics in which the dielectric properties undergo a transition. The two materials have different types of instability, enabling us to build up structures especially tuned for sensing and actuating.

These functional composite materials take advantage of the fact that, when optimizing a smart material, generally one is not attempting to optimize all the electromechanical coupling coefficients, but only those tensor coefficients appearing in the figure-of-merit of the device. Taking advantage of the mixing rules, we build series and parallel connectivity patterns into the composites using an electrically soft ferroelectric phase and a mechanically soft polymer. A variety of transducer materials with large electromechanical coupling coefficients have been constructed in this way. The composite transducers illustrate a very general approach that applies to piezoelectric materials and to many other functional composites.

As we look to the future, many new types of smart systems will be developed. Aerospace engineers are interested in smart airfoils to control drag and turbulence. Diabetics need medical systems to sense sugar levels and deliver insulin. Architects are designing smart buildings that incorporate self-adjusting windows to control the flow of solar energy into houses. Tennis players will want self-adjusting tennis racquets for overhead smashes and delicate drop shots.

Integration and miniaturization of electroceramic sensors and actuators is an ongoing process in the automotive, medical, and consumer electronic areas. Multicomponent, multifuncional ceramics made by tape casting and screen printing are in common use. The next logical step is to combine the sensors and actuators with the control system. This is being done by deposition of electroceramic coatings on integrated circuit silicon chips. A variety of microelectromechanical (MEMS) systems using PZT films on micromachined silicon are under development.

Fiber-Optic Sensors

by Ralph Wastwig, Corning Incorporated

Using fiber optics in a sensor is not a new idea. It dates from the 1960s when optical fibers were used in punched-card optical readers for early computers and in liquid level probes. Shortly after the inventions leading to low-loss glass optical fiber in 1970, the Naval Research Laboratory started a concerted effort to develop a fiber-optic hydrophone. They used phase modulation techniques based on classical configurations, for example, the Mach Zehnder interferometer. The length of single-mode fiber in the reference arm was acoustically insulated, and the sensing fiber was highly coupled to the acoustic pressure field through special transduction coatings. Changes in the optical path length between the two arms due to mechanical strain differences were detected and correlated to the acoustic pressure. Besides long-distance detection of submarines, the highly sensitive optical hydrophones can detect seismic events and the migration of nature's submarine, the whale. Other quantities can be detected applying different transduction coatings. For example, sensitive fiber-optic magnetic sensors were developed by coating the sensing fiber with magnetostrictive materials such as met glass.

Many other techniques have been used in designs of fiber-optic sensors. Although microbending loss is an anathema for telecommunications and much effort was expended in fiber and cable design to minimize it, an early and simple sensor maximized a controlled microbend. Other sensor types include polarization modulation, fiber-grating sensors in which Bragg gratings are permanently formed by an ultraviolet laser, and a variety of designs wherein spectral changes originating at a small device placed in the sensing field are relayed by a fiber to remotely located instrumentation.

Fiber-optic sensors for just about any variable have been developed over the past three decades. They include strain, pressure, vibration, fluid flow, temperature, voltage, current, biomedical and chemical recognition, and rotation rate. Their ultimate success depends on advantages they may offer over conventional methods. Commercialization is most likely for usage in hazardous environments where electrical currents are not permitted, in regions of high electromagnetic fields, and, in cases where the fiber-optic sensor has demonstrated superior sensitivity and dynamic range.

Much effort by both government and industry has gone to the fiber-optic gyroscope, first constructed by V. Vali and R. Shorthill at the University of Utah in 1976. Rotation rate relative to the fixed stars can be determined by measuring the phase change between counterpropagating beams of light around a closed loop—an effect first demonstrated by Georges Sagnac in 1913. Sensitivity is proportional to the area enclosed by the loop. For a loop formed by multiple turns of a long length of single-mode fiber, the sensitivity is increased by a factor equal to the number of turns. The fiber gyroscope and the more mature related technology, the ring-laser gyroscope, are replacing the conventional mechanical gyroscope in inertial guidance systems.

Fiber-optic sensors can be classified as localized or distributed and as extrinsic or intrinsic. An example of a distributed, intrinsic fiber-optic sensor, developed at Corning Incorporated in the 1980s, is based on mode–mode interference in a single length of multimode fiber. The input is coherent light from a laser and the far-field output is a mottled pattern of bright and dark spots. The far-field pattern is detected through a spatial filter. Small mechanical perturbations at any position along the length of the fiber causes phase shifts among the many modes and is registered as a rapidly changing far-field pattern. Applied as an intrusion detector, the sensor is currently installed at many locations including U.S. embassies, nuclear power plants, and along international borders.

Zirconia Oxygen Sensors

by Eleftherios M. Logothetis, Ford Motor Company

To satisfy stringent emissions control regulations, modern automobiles use electronic engine controls, catalytic converters for exhaust-gas treatment, and many sensors. The most important of these sensors is the exhaust-gas oxygen sensor based on ZrO_2. Placed in the exhaust gas of an engine, this sensor controls the air-to-fuel ratio (A/F = mass of air/mass of fuel) of the intake mixture of the engine to be at the stoichiometric value, where the so-called three-way-catalyst (TWC) has its highest efficiency for the elimination of CO, hydrocarbons, and oxides of nitrogen, as shown in Fig. 24.

Without the exhaust-gas oxygen sensor, it is impossible to achieve the low emissions of present and future automotive systems. The usefulness of an oxygen sensor for A/F control results from the fact that, at thermodynamic equilibrium, there is a unique relationship between oxygen partial pressure (P_{O_2}) in the exhaust gas and A/F.

In particular P_{O_2} changes abruptly by many orders of magnitude (>20) at the stoichiometric A/F value.

ZrO_2 doped with yttrium is a highly oxygen-ion-conducting material at elevated temperatures. In the 1960s Westinghouse and other laboratories conducted much R&D on ZrO_2 for high-temperature fuel-cell applications. This work also led to the development of laboratory-type ZrO_2 oxygen sensors. In the 1970s and 1980s the automotive need generated a large R&D effort by several manufacturers to develop inexpensive and durable ZrO_2 devices that could operate in the harsh exhaust-gas environment.

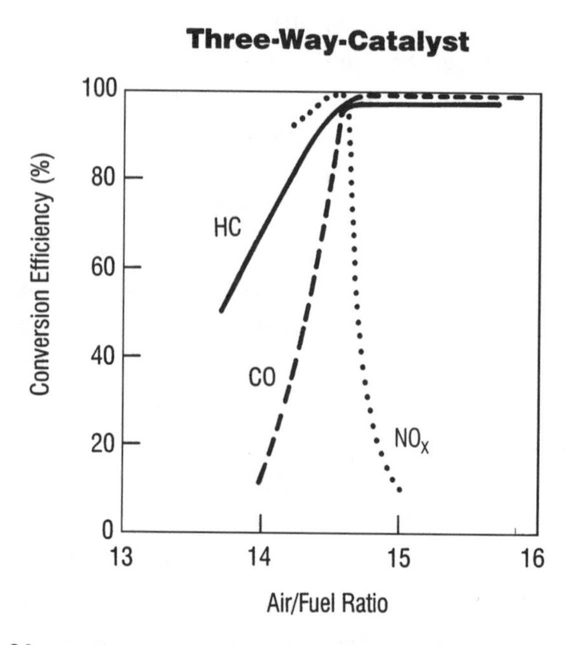

Figure 24. *Conversion efficiency as a function of air-to-fuel ratio. Courtesy of Ford Motor Company.*

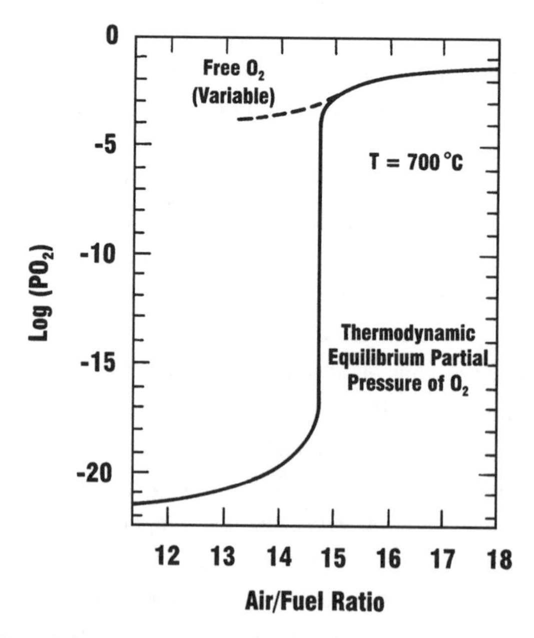

Figure 25. *Oxygen partial pressure as a function of the air-to-fuel ratio. Courtesy of Ford Motor Company.*

A typical automotive oxygen sensor consists of a dense ZrO_2 ceramic in the form of a thimble with porous platinum electrodes on the inner and outer surfaces of the ceramic.

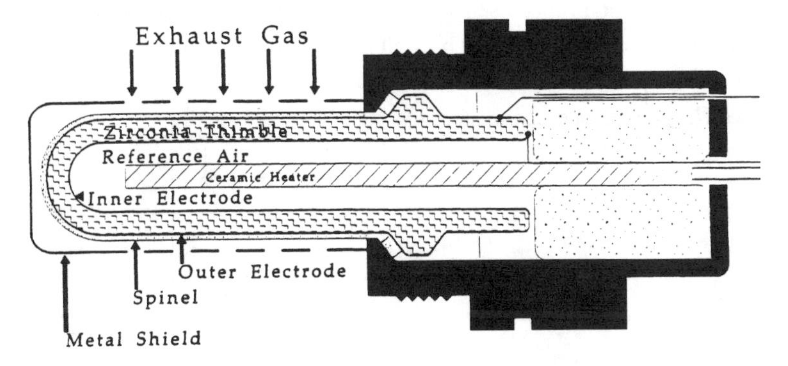

Figure 26. *Section of typical ZrO_2 oxygen sensor. Courtesy of Ford Motor Company.*

The inner chamber is connected to the ambient air as a reference atmosphere and contains a ceramic heater. The outer electrode, which is exposed to the exhaust gas, is covered with a porous inert layer (e.g., spinel) for protection against abrasion and contaminants from the exhaust gas for improved performance. The ZrO_2 sensing element is mounted on a spark-plug-type metal shell. In the exhaust gas, as the A/F ratio (and, correspondingly, the P_{O_2} in the exhaust gas) is changed, the sensor develops a voltage given by the well-known Nernst equation, $V = (RT/4F) \ln (P_{O_2}/P_{O_{2,air}})$. The sensor voltage shows a large stepwise change at the stoichiometric A/F.

A major technological achievement has been the development of materials formulations and ceramic fabrication processes to obtain a ceramic with well-controlled microstructure and containing all three ZrO_2 phases: cubic, tetragonal, and monoclinic. This partially stabilized zirconia (PSZ) has the mechanical and thermal properties needed for the sensor to survive in the exhaust-gas environment for >100,000 miles. A critical part of the technology also is the ability to incorporate good platinum electrodes with the required porosity, catalytic activity (needed because the exhaust gas is not in thermodynamic equilibrium), adhesion, and durability. Tens of millions of these sensors are produced every year.

Recent developments include the use of these sensors also for on-board monitoring of the TWC efficiency, the fabrication of planar structures based on ceramic tapes for improved performance and lower cost, and the development of single- and double-cell ZrO_2 structures operated in the amperometric mode (i.e., passing an electric current through the cell) to obtain more sensitive oxygen sensors and sensors for other exhaust-gas constituents, such as NO_x. ZrO_2 oxygen sensors (of different design) are extensively used in other applications, e.g., control of metallurgical processes in steelmaking.

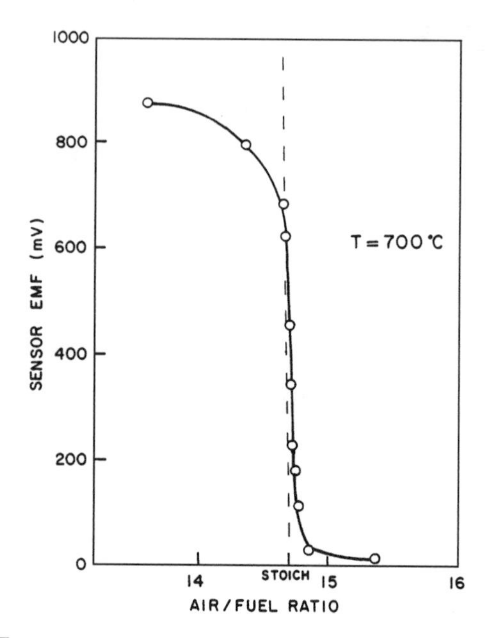

Figure 27. *Sensor voltage as a function of the air-to-fuel-ratio. Courtesy of Ford Motor Company.*

Humidity Sensor for Automatic Microwave Ovens

by Koichi Kugimiya, Matsushita Electric Company

Many materials show variation of resistivity with humidity. However, no materials known before 1978 showed high sensitivity over a wide humidity range and quick response to the change of humidity combined with satisfactory stability or reliability in dirty, real field conditions. Frequent surface cleaning or maintenance was required for materials with large and quick response. A novel and very practical relative humidity sensor was invented by Tsuneharu Nitta of Matsushita using a very stable complex $MgCr_2O_4$–TiO_2 oxide.

The oxide is fired following conventional ceramic processing and tape casting. The resulting oxide has very fine grains and micropores of <1 μm and about 30% volume fraction of pores. It is actually a type of controlled nanostructure material. The change of resistivity is 3–4 orders of magnitude over the relative humidity range of 5%–95%, and the sensor can work up to 500°C at a relative humidity as low as 1%.

A humidity sensor introduced as "Humiceram" has been very successful. This sensor material is tough and operates for more than 5 years in dirty, oily, smoke-filled, and misty kitchen or cook spaces without maintenance. The dirt on the sensor material is burned off regularly to refresh the surface automatically. The burnoff process causes no damage to the material. The first application of the sensor was in an automatic microwave oven that became a best seller in U.S. and Japanese markets. The sensor monitors only moisture in the exhaust of an oven, yet it can precisely define the state of cooking with a given time and input power of an oven for most of the cooking processes. The sensor and control system automatically stop the oven when the food is done, allowing trouble-free cooking for anyone.

Recently Humiceram has been modified, and its applications have been expanded to various uses requiring precision humidity control in the field. For example, it is used for precision humidity control of room atmosphere for sensitive equipment, such as printing machines.

Structural Ceramics

Ceramic Armor

by S. R. Skaggs, Los Alamos National Laboratory, retired, and Jack A. Rubin, consultant

Ceramic armor was developed beginning in the 1950s to provide improved protection at a practical weight and cost. Traditional armor was made of steel but became heavier as bullets moved from lead to steel to hardened steel to tungsten and to metal-bonded tungsten carbide. An armor at least as hard as the bullet is needed. Other factors, such as toughness via improved microstructure and constraint on motion of fractured armor fragments by strongly confining it also are important.

The first ceramic armor in the 1950s was an 85% alumina (soon upgraded to 99.5% alumina) used for body armor and seats for helicopter pilots and gun shields for gunboats. Subsequently, hot-pressed boron carbide was used, especially for helicopter seats. A reaction-bonded boron carbide ceramic (with a silicon carbide bond) was then developed to provide a lightweight body armor that could be formed to shape and-was economic. The major ceramic armor materials today are alumina, silicon carbide, titanium diboride, boron carbide, and aluminum nitride. Organizations involved in the development of ceramic armor include Carborundum Company, Ceradyne, CerCom, Coors Porcelain Company, DARPA, Dow Chemical Company, Eagle-Pitcher, GTE Sylvania Materials Laboratory, Lanxide Armor Products Company, Livermore National Laboratory, Los Alamos National Laboratory, Morgan Matroc, and the U.S. Army Research Laboratory.

The mass effectiveness of ceramic armor is defined as the mass per unit area of heat-treated 4340 steel (RHA) needed to stop a given projectile divided by the same quantity for ceramic armor for that projectile; it is used as a measure of performance. The mass effectiveness value ranges from ~1.2 for alumina to as much as 2 for titanium diboride and silicon carbide.

Performance of ceramic armor is enhanced by containing and confining it in a material that holds it in compression to prevent escape of shattered fragments. These fragments then erode or grind down the penetrator. The penetration hole in a confined ceramic typically is only slightly larger than the penetrator. The tip of the penetrator erodes to a pencil shape, breaks off, and the process repeats.

Attempts to improve toughness of ceramics led in the 1980s to the development of ceramic–metal composite armor materials, including silicon carbide–aluminum and boron carbide–aluminum. A combination of metals, ceramics, and cermets now are being used in light armors. Steel continues to have a place, principally in heavy armors.

Ceramic armor has fairly wide use on armored personnel carriers to increase their capability, especially against artillery shrapnel. At the thicknesses used it provides complete protection against 0.30 caliber armor-piercing bullets and very good protection against 0.50 caliber. Silicon carbide bonded with aluminum is used in C-141 and C-17 cockpits and as part of the armor covering on armored personnel carriers. The

Marine Corps LAV-25 vehicles use it when battle conditions dictate. Boron carbide seats are primarily used in combat helicopters. Alumina ceramics are used on the German Leopard tank skirts and on the deckhouse of the landing craft air cushion (LCAV) vehicle by the Marine Corps. The most widespread use of ceramic armor is in bullet-proof vests. Many armies and police forces use them.

Research on ceramic armor is continuing. Several scientists have studied natural ceramic structures and have attempted to formulate armors simulating such things as oyster shells with harder materials. These have been given the name biomimetic materials. Using the model that nature created, others have attempted to make functionally gradient materials in similar layered fashion with the hard material on the face that meets the penetrator and graded to a tougher material as the backup. Sometimes as many as 200 layers are used.

Continuous Ceramic Fibers

by Richard E. Tressler, The Pennsylvania State University

The development of continuous ceramic fibers has been primarily stimulated by the demand for higher-performance continuous fibers for fiber-reinforced polymers, metals, and ceramics. There also is a niche market for ceramic-fiber textiles for high-temperature resistant curtains, sleeves for thermocouples, and oil spill cleanup booms (upon which the crude oil is burned). Glass fibers and carbon fibers developed the composite markets, and the demand has grown for higher modulus-to-density- and higher strength-to-density-ratio materials and for fibers with high-temperature strength and creep and oxidation resistance. Several developments have contributed to this major innovation of continuous ceramic fibers.

In the 1960s the military market demanded high-performance fibers for reinforcing polymers and aluminum alloys. The U.S. Air Force spearheaded the development of chemically vapor deposited (CVD) deposition of monofilamentary boron on a tungsten substrate. The CVD deposition of monofilaments commercialized by now Textron (then AVCO) is a mature technology with a limited market for boron fibers (aerospace and high-performance recreational gear) and an emerging market for SiC (deposited on carbon) fibers.

The strong interest in the development of ceramic-matrix composites has stimulated major innovation in the development of small-diameter, weavable, continuous ceramic fibers. Two major developments are noteworthy: the use of sol–gel precursors to form oxide fibers and the use of silicon-containing polymers as precursors to form non-oxide fibers. The first one was pioneered by 3M Company (H. Sowman, also discussed as an innovation). 3M's line of fiber products carry the trademark Nextel™. The specific fibers range from Al_2O_3–SiO_2–B_2O_3 fibers (Nextel 3120), which contain a mullite phase and an amorphous phase, α- Al_2O_3 fibers (Nextel 610), which were developed to reinforce metals, to Al_2O_3–SiO_2 high-temperature fibers (Nextel 720), which contain both α- Al_2O_3 and mullite. The preceramic polymer route was pioneered by Yajima in Japan with a process in which a polycarbosilane was crosslinked and then pyrolyzed to a small-diameter fiber consisting of nanocrystalline β-SiC, turbostratic carbon, and an amorphous Si-O-C phase. Nippon Carbon commercialized this development under the trademark Nicalon™ fibers. They now produce a family of fibers based on this approach, including a nearly oxygen-free, nearly stoichiometric β-SiC fibers (Nicalon S). Other companies have pursued similar ceramic processing routes to form non-oxide fibers. Most notable among these companies is Ube, which markets a line of Tyranno™ fibers. The Bayer Company is developing an apparently amorphous Si-B-C-N-O fiber with reportedly excellent high-temperature properties, but is not commercially available at this writing. The Dow Corning Company is commercializing a product under the trademark Sylramic, which is a polymer-derived fiber but has been completely pyrolized and densified to a stoichiometric β-SiC polycrystalline fiber (doped with titanium and boron), which has excellent high-temperature load-bearing capabilities.

The innovations in continuous ceramic fibers have been of two types: (1) the development of large diameter (>100 μm) monofilaments using CVD deposition on a small-diameter substrate fiber (boron and SiC are the major materials now available commercially) and (2) the development of small-diameter (10–15 μm), weavable fibers in the form of tows of fibers using sol–gel precursors for the oxides and preceramic polymers for the non-oxides.

Tough SiC-Based Thermostructural Ceramic-Matrix Composites

by Roger R. Naslain, University of Bordeaux, and Pierre J. Lamicq, SEP Division de SNECMA

The use of monolithic SiC-based ceramics in thermostructural applications, e.g., as in components of engines, has been limited up to recently by their brittleness, an intrinsic property of bulk ceramics. It was discovered (in the late 1970s) that SiC-based ceramic-matrix composites (CMCs), comprising a SiC matrix reinforced with continuous ceramic fibers, such as carbon or SiC fibers, can be tough materials when the fiber–matrix bonding is *weak enough.* The concept of CMCs with weak fiber–matrix interfaces is one of the major breakthroughs of the century in the field of structural ceramics inasmuch as its use opens new fields of applications for ceramics in severe environments.

Contrary to most polymer- and metal-matrix composites, CMCs are inverse and damageable composites. Under load, the brittle matrix fails first, the matrix microcracks being deflected in the fiber–matrix interfaces, at least when the fiber–matrix bonding is weak enough.

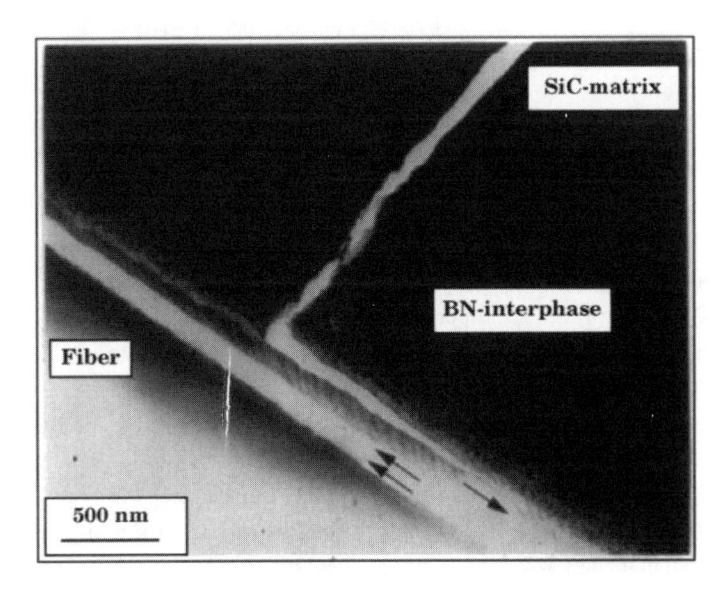

Figure 28. *Matrix crack deflected along fiber interface. Courtesy of University of Bordeaux and SNECMA.*

Thus, the interface acts as a fuse, protecting the brittle fibers from early failure. As a result, the load can be progressively increased with simultaneous fragmentation of the matrix, debonding of the fibers over a certain length, and sliding of the debonded fibers.

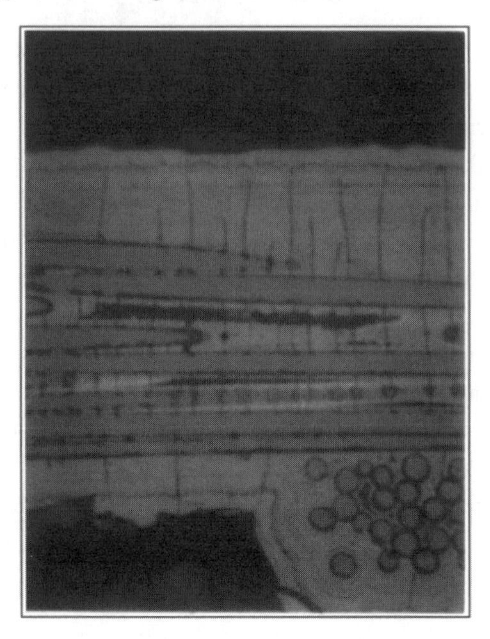

Figure 29. *Multiple matrix cracking. Courtesy of University of Bordeaux and SNECMA.*

These damaging phenomena are responsible for a nonlinear stress–strain behavior, an unusual feature for ceramics. When finally the fibers fail, their pullout from the matrix continues to require a large amount of energy. All these phenomena account for the toughness and reliability of SiC-based CMCs with weak fiber–matrix interfaces.

The control of the fiber–matrix bonding in SiC-based CMCs is achieved through the use of a submicrometer layer of a material displaying a low shear strength that is deposited on the fiber surface and referred to as the interphase. Suitable interphase materials are preferentially those with a layered crystal structure (pyrocarbon or boron nitride) or a layered microstructure, such as multilayers of pyrocarbon-SiC. Although SiC-based CMCs can be processed according to various procedures, it was shown in the late 1970s by R. Naslain and his co-workers and by E. Fitzer and his co-workers that a gas-phase route now referred to as the CVI process (for chemical vapor infiltration) is particularly suitable. In this process, the interphase and the SiC-matrix are successively deposited on the fiber surface within a porous fiber preform from gaseous precursors (e.g., a hydrocarbon for pyrocarbon and CH_3SiC_{13}/H_2 for SiC) at relatively low temperatures and under reduced pressures, the process being highly flexible but relatively slow when conducted under isothermal/isobaric conditions (I-CVI). The densification process can be accelerated by applying to the fiber preform a temperature/pressure gradient, as in the F-CVI process developed by D. P. Stinton and his co-workers in the mid 1980s. Finally, the interphase can be engineered at the nanometer level through the use of pressure-pulsed CVI (P-CVI), as very recently established by R. Pailler and his co-workers. Thus, CVI under its different versions is a unique process to produce SiC-based CMCs.

At the industrial level, SiC-based CMCs production has been developed by Société Européene de Propulsion, France, now SEP, Division de SNECMA. The process was licensed to E. I. du Pont de Nemours and Company, then DLC in Germany. A variety of parts and high-temperature structures have been processed and tested at scale 1 for many uses: rocket motors and exit cones, very hot gas valves and pipes, ramjet chambers, turbojet engines, and atmospheric reentry vehicles. The first series production deliveries of a SiC-based CMC part started early in 1997 with C/SiC nozzle petals of the SNECMA M88 engine. Large parts, ~2 m long, have been generated for the space plane.

SiC-based CMCs produced by SEP, from oxygen-free SiC fibers to oxidation-resistant interphase and self-healing matrices, already display lifetimes of the order of 1000 h under cyclic loading in air at high temperatures. Further improvements are anticipated, as the basic concept of nonbrittle SiC-based CMC can be extrapolated to other carbon and ceramic composites. It is foreseen that the fields of application for these materials will continue to be enlarged, provided further progress is made at the level of the fibers and the interphases.

Melt-Infiltrated SiC Ceramics and Composites

by Mrityunjay Singh, NASA Lewis Research Center

Melt-infiltration processing of silicon carbide ceramics started in the second half of the 20th century in Europe, the United States, and other parts of the world. Popper and his co-workers (1960) from the U.K. Atomic Energy Authority (UKAEA) have reported the microstructure and properties of reaction-bonded silicon carbide. Various processing research and development activities were conducted at many universities and institutes, at Carborundum Company and Norton Company, along with other ceramic developers and manufacturers worldwide. These materials were fabricated by silicon infiltration of silicon carbide and carbon-containing green bodies (preforms). The resulting materials, commonly called reaction-bonded silicon carbide (RBSC), contained silicon carbide grains bonded with a silicon phase. There was limited control over the microstructure and second-phase distribution. In an effort to control the microstructure and properties of melt-infiltrated silicon carbide ceramics, many approaches were used to fabricate preforms. They include the incorporation of carbon fibers in the preforms as well as fabrication of microporous carbon preforms by the phase-separation process. Reaction-bonded silicon carbide ceramics are one of the most widely used engineering ceramics worldwide in aerospace, energy, electronics, automotive, process, chemical, and environmental industries.

Fiber-reinforced silicon carbide-based ceramic composites have potential applications in aeronautics, energy, electronics, nuclear, and transportation industries. In aeronautics, these materials are being considered for applications in hot sections of jet engines and nose cones and leading edges of reentry vehicles and hypersonic aircraft. Applications in the energy industries include radiant heater tubes, heat exchangers, heat recuperators, gas filters, and components for land-based turbines for power generation as well as for various components of fusion reactors. Despite their great performance potential, the affordability of these composites has been a major concern in their wide-scale applications.

The utilization of the melt-infiltration process for composite fabrication started in the 1980s with research activities at NASA Lewis Research Center, Cleveland, OH, General Electric Company and DLR in Cologne, Germany. At NASA Lewis, the reactive-melt infiltration technique has been used to fabricate silicon carbide-based ceramic composites reinforced with a variety of commercially available fibers. The technique reported here requires fabrication times significantly less than other processing techniques, such as chemical vapor infiltration (CVI) or polymer infiltration and pyrolysis (PIP). The products produced by this process are nearly fully dense, with good control of microstructure, second phases, and mechanical properties.

In the first step of the composite fabrication process, a fiber preform is fabricated and interfacial coatings (single or duplex) are applied to provide toughening through debonding and crack deflection and prevent molten silicon attack on fibers. There is no limitation on shape, size, and type of fiber preforms. It could be two- or three-dimensional woven or braided preforms of carbon, silicon carbide, alumina, or other types of fibers. The interfacial coating can be applied by vapor- or liquid-processing routes, and its composition and thickness depend on the type of the fibers and processing conditions. In the second step, a low-cost resin mixture is infiltrated into the fiber preforms to net shape and pyrolyzed to give an interconnected network of microporous carbon in the precursor matrix. The chemistry and morphology of the preform is critical in controlling the infiltration kinetics and control of secondary phases in the matrix. These preforms are easily machinable and handleable, because they have good strength. In the third and final step of the process, these preforms are infiltrated with molten silicon or a silicon-refractory metal alloy for 20–30 min to convert the carbon precursor matrix to dense silicon carbide with tailorable microstructure and properties. These tailorable matrices are highly desirable for the improvement of mechanical properties (fracture toughness, first matrix cracking strain, creep resistance, etc.), thermal properties (thermal conductivity, thermal expansion, thermal shock resistance), and environmental resistance of ceramic composites.

An alternative approach involves slurry infiltration of the fiber preform with silicon carbide powder and melt infiltration. This approach yields a high amount of free silicon in the matrix, and there is no continuous network of silicon carbide phase in the matrix. These features are detrimental to high-temperature themomechanical properties and limit the temperature capability of composites.

Transformation Toughening in ZrO_2-Based Ceramics

by Arthur H. Heuer, Case Western Reserve University

Improving the strength, toughness, and overall structural reliability of ceramics has been a nearly constant goal of ceramists for the past century. This has been particularly true in the past 50 years, when the potential of ceramics as light, stiff refractory materials offered the possibility of significantly improving the performance of advanced engineering systems.

A significant advance was initiated by the discovery of transformation toughening in ZrO_2-based ceramics in the mid-1970s. This discovery was made initially by Garvie and his colleagues in Australia, and was followed quickly by work by Heuer in the United States, both groups working on what are now known as partially stabilized zirconias or PSZs. Claussen in Germany had similar interests in microcrack toughening in ZrO_2-based ceramics in the same time frame, and he and Rühle subsequently showed that transformation toughening could be induced in ceramics containing ZrO_2 as a dispersed phase (ZrO_2-toughened ceramics or ZTCs). Work by Gupta in the United States led to the third major class of transformation-toughened ceramics, tetragonal ZrO_2 polycyrstals (TZPs); Gupta's work was advanced by Lange in the United States and by a host of Japanese investigators. The phenomenon of transformation toughening involves the well-known polymorphism of ZrO_2. Pure ZrO_2 crystallizes from its melt at ~2710°C as a cubic phase with the fluorite (CaF_2) structure. In the pure or unstabilized form, ZrO_2 undergoes a displacive but nonmartensitic phase transformation to a tetragonally distorted version of the flourite structure at ~2280°C. A further reduction to monoclinic symmetry occurs via a martensitic transformation; the martensite start temperature is size dependent but occurs at ~1200°C in bulk crystals. This latter transformation involves a volume increase of ~3% on cooling and had been a bane to ceramists wanting to exploit the otherwise excellent refractory properties of ZrO_2 (high melting point and low thermal conductivity). The anisotropic shape strains associated with this volume change caused microcracking and even disintegration of undoped polycrystallized ZrO_2 ceramics.

Research in the 1950s by Hund and Wagner in Germany and Kingery in the United States (following very early work by Nernst in Germany at the turn of the 20th century) that "alloying" ZrO_2 with certain bivalent and trivalent additives (MgO, CaO, Y_2O_3, etc.) "stabilized" the cubic phase against transformation. Such stabilized ZrO_2 had very high ionic conductivities because of the generation of very high densities of charge-compensating oxygen-ion vacancies (one per magnesium or calcium ion, for example), and useful properties as oxygen sensors, solid-oxide fuel cells and the like. However, these materials had very poor

Figure 30. *Martensically transformed ZrO_2 precipitates in partially stabilized zirconia. Courtesy of Arthur H. Heuer.*

thermal shock resistance, because of a combination of high thermal expansion coefficient and very low thermal conductivity. "Partially stabilized" ZrO_2, a mixture of the cubic and monoclinic phases, was produced by reducing the amount of stabilizer dopants added, and significantly improved the thermal shock resistance compared to fully stabilized materials.

The mechanism of transformation toughening in ZrO_2 involves two distinct phenomena, both related to the difficulty of nucleating the martensitic transformation. First, the martensite start temperature of tetragonal ZrO_2 can be lowered to below room temperature by controlling the stabilizer concentration in this phase, and, even more importantly, the martensite start temperature of the particle or grain size of this phase decreases with decreasing size. Second, however, such fine particles undergo the martensitic transformation in the stress field of propagating cracks. The volume change resulting from the stress-induced transformation in these fine tetragonal ZrO_2 particles in a zone around the crack creates favorable tractions on the crack faces, which impedes further crack propagation and significantly increases the fracture toughness (to values comparable to cast iron). The several types of transformation-toughened ceramics represent different strategies for forming transformable ZrO_2 particles in an appropriate matrix. Aside from the variety of new ceramics now being exploited for their good erosion, corrosion, and thermal resistance, understanding the phenomena of transformation toughening led to important applied mechanics insights that were applicable to a range of tough ceramics, including ceramic composites. They also provided additional motivation for chemical companies to enter the field and develop a range of sinterable powders of controlled composition and easy processability. The broader principle of using two-phase materials, in which the external fields can drive transformation(s) to improve material responses is finding applications in other areas, such as ferroelectrics.

High-Performance Reactive Powder Concrete

by Philippe R. Boch, University Pierre and Marie Curie

Much research work has been done over the years to develop high-performance concretes. Two ways recently investigated are compact granular matrix concretes with high superplasticizer and microsilica content and incorporating hard aggregates, and polymer pastes mixed with aluminous cements (MDF, i.e., macro-defect-free cements). Moreover, fiber reinforcement can bring some ductility. However, the point is to make industrial materials that combine high mechanical characteristics with reasonable workability and good durability in various environments and are economically competitive. These rather contradictory requirements are now fulfilled by the reactive powder concretes (RPCs) developed in the 1990s by Richard and Cheyrezy (Bouygues, France).

The RPC program has taken into account five principles: enhancement of homogeneity, increase in density, optimization of microstructure and phase composition, enhancement of ductility, and maintaining fabrication procedure close to existing practice. In RPCs, the binder is made of portland cement mixed with microsilica (microsilica/cement ratio of ~0.25) deflocculated with polyacrylate-based superplasticizers. Microsilica is an essential ingredient to favor the pozzolanic reaction that allows reduction of portlandite and formation of C-S-H with a composition close to tobermorite. Crushed quartz also plays a role. Fine sand (150–600 μm) replaces coarse aggregates, and the paste/aggregate ratio is higher than in conventional concretes. Optimization of the granular mixture and use of very low water/binder ratios (~10%) allow relative densities to be close to 90% and pore diameters to be <5 nm. Heat treatments at 90°C under normal pressure accelerate the pozzolanic reaction and modify the microstructure of the hydrates; if conducted at high temperatures (up to 400°C), they lead to the formation of crystalline xonotlite and are accompanied by major dehydration of the hardened paste. Finally, incorporation of 2 vol% of fine (0.15 mm $\times$ 13 mm) steel fibers dramatically increases fracture energy (from 30 to 20,000 $J{\cdot}m^{-2}$) and allows ductility (ultimate elongation of 5–7 $\times$ 10^{-3} m/m). Such a ductile behavior also can be obtained by confinement in metallic tubes. RPC200 (compressive strength of 200 MPa) involves only 20°–90°C heat treatment. RPC800 (compressive strength as high as 800 MPa) requires 50 MPa presetting pressurization and 250°–400°C heat treatment.

Production and application of RPC200 are close to those of conventional high-performance concrete. By reason of high ductility, the material can be used in prestressed structures not incorporating passive steel reinforcement. Such structures can be ~3 times lighter than equivalent structures made of conventional concrete. RPC800 can be used only for production of precast elements. Its exceptional resistance to impact suggests its use for military structures. Dense microstructures of RPCs give structures advantageous waterproofing and durability characteristics, of special interest for industrial and nuclear waste storage.

The first (1997) spectacular example of an industrial structure made of RPC is the pedestrian/bikeway bridge that spans the Magog River in Sherbrooke (Quebec). For a span of 60 m, the slab thickness is only 30 mm. RPC development has followed the way paved by ceramists, of microstructure optimization (porosity reduction and microcrack elimination), and quest for high toughness (use of reinforcing fibers). RPCs really deserve to be considered as high-performance, chemically bonded ceramic materials.

Chemical Tempering of Glass Products

by Emil W. Deeg, AMP, Inc., retired

Surface flaws acting as stress raisers cause glass products to have a tensile strength far below the high intrinsic strength of glass as a material. They must be eliminated or at least rendered ineffective relative to stress levels to which specific glass articles are expected to be exposed. Removing a thick enough surface layer by nonmechanical treatment can eliminate flaws temporarily. For silicate glasses this is done by etching with aqueous solutions containing F^- ions. However, depending on environmental conditions an article is exposed to, the so-achieved strength increase is short-lived.

Methods for rendering strength-reducing surface flaws ineffective use the concept of introducing conditions in a product that result in compression of its surface so that the effect of stress raisers is compensated. The most common technique is controlled quenching of a hot glass article or portions of it. Because no change in glass composition is required, it is sometimes referred to as physical tempering. The oldest technique for toughening a glass article by changing the composition of the surface relative to the interior consists of fusing layers of low-expansion glass onto opposite sides of an article made of a higher expansion glass. Recent techniques include selective volatilization of surface components resulting in low-expansion layers and modification of the composition of the surface by exchanging glass constituents so that the thermal expansion of the surface layers is reduced. The latter approach became known as ion-exchange toughening, ion stuffing, or by the less-specific terms chemical strengthening and chemtempering. This technique is performed by exposing glass articles at elevated temperature to salt pastes applied to the surfaces or by submersing them in salt melts. Increased durability of glass articles resulting from certain treatments at the hot end of container glass lehrs can be understood as ion-exchange toughening in the vapor phase. Ion-exchange toughening by formation of low-expansion, crystalline surface layers has been demonstrated on a laboratory scale.

The outstanding advantage offered by ion-exchange toughening is that the degree of surface compression is essentially independent of the thickness of an article and can be made very high compared to that achievable by air quenching. Thin-wall glass articles and articles with varying wall thickness can be ion-exchange toughened. The stress distribution $F(x)$ in a toughened sheet of glass of thickness, d, can be described by the empirical expression

$$F(x) = bx^{2n} - a$$

where $-d/2 \leq x \leq d/2$, a the maximum tension at the center of the sheet, and b the measure for the maximum compression at the surface. The exponent $2n$ is a measure for the slope of $F(x)$. It depends on glass composition and treatment parameters. The higher the value of n, the steeper the slope of the stress distribution. For air-quenched, flat articles, $n \approx 1$; i.e., the stress distribution is approximately parabolic. For ion-exchange-toughened articles, $n \approx 3$ to $n \approx 4$, or even higher. The thickness of the compression layer t becomes

$$t \approx |(a/b)^{1/2n}|$$

which, considering the values of n, shows the low thickness of the compression layer for an ion-exchange-toughened sheet relative to that of the air-quenched product.

Besides hot-end lehr treatments, the most frequently used ion-exchange-toughening process consists of submersing glass articles in a KNO_3 melt. Originally, the method was applied to soda–lime–silica glasses and resulted in an exchange of Na^+ by K^+ ions and was restricted to temperatures below the strain point of the glass. Soon it was extended to other glass compositions, particularly ophthalmic glasses. The feature to toughen glass articles with nonuniform thickness made ion-exchange toughening most attractive to eyeglass manufacturers. A process that would allow, without adjusting process parameters, toughening of strong positive lenses (e.g., +5 diopters) with high center thickness as well as strong negative lenses (e.g., −10 diopters) with very low center thickness was very desirable.

Treatment below the strain point required a 12–16 h long exposure in the KNO_3 melt to toughen lenses to acceptable impact resistance and thickness of the compression layer. This processing time was unacceptable from a cost and customer service viewpoint. Despite the technical advantages ion-exchange toughening offered, air quenching of safety and prescription lenses was preferred mainly because of the established throughput rates of minutes per lens. The situation changed when, after a critical evaluation of the definition of the strain point, it was shown that crown glass lenses could be ion-exchange toughened to ANSI Z-80 impact resistance if treated above the strain point. Optimized batch processing now permitted next-day delivery of prescription eyeglasses. An additional advantage was that, except for treatment of striking glasses (e.g., Cd-S-Se-, Cu-, Ag-, Au-"rubies", photochromics), ion-exchange toughening required less rigorous process control than air quenching.

High strength of a glass product is very often a secondary performance characteristic. Therefore, optimization of glass compositions for ion exchange has to retain or, if economically acceptable, improve the primary product characteristics. Lithia–phosphate–alumina glasses used today for high-power neodymium-doped glass lasers offer the desirable primary performance, i.e., laser action, and are suitable for toughening laser rods or disks by a K^+ for Li^+ ion exchange. The situation is more complex, e.g., for chemical apparatus glass articles or for ophthalmic products that are manufactured on a large scale. Ophthalmic crown glass compositions may serve as example. For optical reasons the refractive index at the He *d* or Na *D* line and the Abbe number < *d* or < *D* must be retained. The index requirement also is necessary to avoid very costly tooling changes at the manufacturing level. To permit production of fused bifocal lenses without costly modification of established processes and development of new high-index segment glasses, thermal expansion and viscosity versus temperature functions of the crown glass must be maintained.

In-Situ-Reinforced Silicon Nitride

by Chien-Wei Li, Allied Signal, Inc.

In the late 1970s the U.S. Department of Energy started an automotive gas turbine program as a result of the oil embargoes, and it determined that the only way to meet the new fuel and emission requirements of a standard car would be through the use of ceramics in the hot section of the turbine engine. It was envisioned that turbine blades and nozzles in the future engines would be made of lightweight, high-strength, high-toughness, and oxidation-resistant silicon nitride operating under uncooled conditions at temperatures where superalloys would have already melted. With an aggressive 1370°C turbine inlet temperature goal, the ceramic engine would run much more efficiently than traditional engines. This revolutionary approach would create a new paradigm in engine technologies and impact the automotive and the aerospace and electric power industries.

Silicon nitride is a covalently bonded ceramic that can be densified through liquid-phase sintering. In the 1970s only a limited number of silicon nitride materials were available, and those had properties far below those required for turbine engine applications. Namely, their fracture toughness was low (4–5 MPa·m$^{1/2}$), strength variation was high (Weibull modulus <10), and their temperature capability was limited to ~1200°C because of the softening of grain-boundary second phases formed by low-melting sintering aids. Limited by the powder quality, these materials could be densified only by hot pressing, a process not amenable to cost-effective production of complex-shaped parts.

Over the past three decades, however, tremendous progress has been made in developing better powders, net-shape forming technologies, low-cost sintering processes, optimum microstructure, and refractory compositions. Today, silicon nitride materials have been fabricated as turbine blades, nozzles, seals, cylinder engine valves, and turbochargers and successfully used in auxiliary power units, industrial turbine engines, propulsion engines (seals), and automotive engines. The development of *in-situ*-reinforced (ISR) silicon nitride was driven by the need of a high-toughness ceramic for these applications. The ISR material has a microstructure consisting of elongated hexagonal β-Si_3N_4 grains developed *in situ* during sintering. During liquid-phase sintering, α-Si_3N_4 dissolves in the liquid formed by molten grain-boundary oxide/nitride glasses and reprecipitates on the existing β-Si_3N_4 nuclei present in the starting powder. The β-Si_3N_4 grows to rod-shaped grains, because of faster growth in the *c*-axis direction and the maximization of its low-energy prism planes. In the fully dense article, these elongated grains act similar to whisker reinforcement and improve the toughness and strength properties of the ISR silicon nitride as was first noted by Lange in 1973. The manufacturing of engine-quality silicon nitride parts with high fracture toughness became a reality with gas-pressure-sintering technology, which enabled the sintering of silicon nitride at temperatures >1800°C by repressing the dissociation of silicon and nitrogen through the use of overpressures of nitrogen gas. In 1989, using a multiple-step gas-pressure-sintering process, Li and Yamanis made an ISR silicon nitride having a large elongated-grain structure and a fracture toughness value of ~11 MPa·m$^{1/2}$. The material had a pronounced *R*-curve behavior or a fracture toughness value that increased with the crack size. Moreover, it showed an unusually high thermal conductivity value of 87 W/(m·K), or ~50% higher than that of other silicon nitrides at that time.

The increased toughness with crack size behavior is the result of bridging by debonded elongated grains across the crack surfaces. As the crack extends in the ISR material and crack surface separation increases, the bridging grains elastically deform and pull out, thus increasing the fracture toughness. Because of its *R*-curve, an ISR silicon nitride typically shows high Weibull modulus (>20), damage tolerance, and fatigue-resistant properties important to structural applications. In the 1990s there has been a burst of research on the correlation between the grain size, grain-size distribution, and grain aspect ratio of silicon nitride on crack-bridging behavior and *R*-curve behavior. It has been recognized that there exists an optimum grain size and grain-size distribution for the best combination of mechanical properties. The achievement of this condition has been the focus of research in this decade. Through sintering aid selection, starting powder α/β phase variation, and sintering process manipulation, one can tailor the resultant microstructure of an ISR silicon nitride. Combined with the understanding of micromechanical fracture mechanisms and the roles of different microstructural elements, it is possible to design the best ISR silicon nitride for any specific application. Finally, the focus of development work in this area in the next century will be to deliver cost-effective ISR silicon nitride parts with superior and dependable properties so we can truly benefit from this elongated-grain structure that nature has given us.

Silicon Nitride Turbocharger Rotors

by John Holowczak, United Technologies, and Terry Sakamoto and David Carruthers, Kyocera, Inc.

In 1985 the ceramic engineering community was facing a considerable challenge. Using small gas turbine engines for automobiles was widely forecast to be feasible, provided that the components used in the hot section could be produced from high-strength, high-temperature-resistance ceramic materials, such as silicon nitride and silicon carbide. Methods of reliably producing these materials, in complex form, such as turbine rotors, had yet to be proved. Properties achieved in simple plates and test bars could not be reliably reproduced in complex-shaped components.

Turbocharger rotors were identified as an application that could make use of structural ceramics having lower density and higher heat resistance than metal alloys. Turbochargers are used to compress the air entering piston engines, to provide greater horsepower and torque from a given engine displacement. Typically, turbocharger rotors are produced from Inconel superalloys. Silicon nitride offered the possibility of sharply lowering the turbocharger's moment of inertia. This in turn would allow reduction of the hesitation common to turbocharged cars of that era due to the time required for the turbocharger to spin up to high speed. For a silicon nitride turbocharger rotor, steady-state tensile stresses can approach 280 MPa (40 ksi). Lastly, the complexity of a turbocharger is similar to that of a turbine for a turbine engine.

In 1987 the Nissan Fairlady Z (the Japanese version of the Nissan 300Z sold in the United States) was introduced with twin ceramic turbochargers, made by NGK via powder injection molding. Toyota soon followed, with ceramic turbocharger rotors produced by Kyocera using their Hybrid Molding process, which is similar to wax-based injection molding. Because of the high taxes assessed for large-engine-displacement automobiles in Japan, turbocharging was quite popular, and production of ceramic-based turbochargers reached levels of ~50,000 parts per month by 1991. All of these automotive ceramic turbocharger rotors are proof spun to ensure quality, and failure rates at both NGK and Kyocera were <1 part per month. The application was highly successful in field use

More importantly, the price of silicon nitride turbocharger rotors dropped to a level nearly equal to that of Inconel 718 rotors. This provided proof that ceramics could be cost competitive with metals while performing in a demanding application at temperatures of up to 900°C (1650°F).

First Man-Rated Spacecraft Application for Silicon Nitride Ceramic Bearings

by John Holowczak and Roger Bursey Jr., United Technologies Corporation

Imagine a ball-bearing assembly that must operate at temperatures of −250°F to −350°F (−157°C to −212°C) in an environment consisting of liquid and boiling oxygen. Imagine the bearing was part of a turbomachinery assembly with a shaft horsepower rating of over 25,000.

Hybrid ceramic bearings, consisting of silicon nitride balls in metal races, have been in use in the liquid oxygen (LO2) pump of the space shuttle main engine since July 1995. The pump channels roughly 7400 gallons of liquid oxygen per minute during the ~8 min long space shuttle lift off and flight to orbit.

Prior to 1995 the pumps used all-metal bearings that were part of necessary tear down, inspection, and/or replacement following every couple of space shuttle missions. The ceramic rolling elements were one of the key improvements required to extend the life of the pumps to 60 missions with an inspection interval after every 10 missions. Ceramic hybrid bearings, both ball and rollers, are currently under engine ground testing in the space shuttle liquid-hydrogen fuel pump, at temperatures lower than –350°F.

The hybrid ceramic bearings last longer because the silicon nitride ceramic balls are lighter, harder, and less chemically reactive than the metal balls they replace. The balls thus have a lower rotational load on the outer raceway while spinning. The coefficient of friction between metal and ceramic is much lower than metal-on-metal, thereby reducing bearing temperatures.

The ceramic hybrid LO2 bearing was developed by United Technologies Pratt & Whitney, under a contract from NASA to build a more durable oxygen pump. The silicon nitride rough ball blanks were manufactured by Toshiba Corporation and supplied by Enceratec, Inc. FAG Bearing Company finished the balls and assembled them into bearings. Ceramic engineers from United Technologies Research Center assisted their Pratt & Whitney Division in defining methods for assessing the quality of the silicon nitride balls.

6

Special Applications of Ceramics

This section presents ceramic innovations engendered by the need for materials for many special applications. Medical needs have led to the development of a variety of bioceramics, including bioactive glasses and hydroxyapatite coatings to assist bone growth. A ceramic innovation that directly impacts almost everyone is dental restorations based on ceramic-fused-to-metal and the subsequent development of strong, all-ceramic restorations. Ceramics are important in medical instrumentation also. A particular case is the ceramic scintillator developed for X-ray detectors in CT-body scanners.

Nuclear ceramics include uranium dioxide fuel for reactors and plutonium-based power supplies for unmanned spacecraft. Special glasses for nuclear waste immobilization are a key part of the national plan for dealing with radioactive waste.

Environmental uses of ceramics include the very important technology of automobile exhaust cleanup based on catalysts carried on honeycomb ceramics.

Other special uses of ceramics include ceramic filters with ultrafine holes. These filters appear to have a bright future in filtration of liquids and gases. Ceramic filters for water purification appear to be finding a niche market.

Phase diagrams for ceramists is presented here as an innovation in research and organization of results in useful form that is basic to the development of many new ceramics.

Thermoluminescence dating of ceramics is a new technology important to archaeology and to museums.

An innovation familiar to everyone but not generally recognized as involving ceramics is the heating element used in electric stoves. A tubular layer of ceramic is a concealed but essential part.

Another innovation universally familiar but usually unrecognized as involving ceramics is paper. A small amount of ceramic filler gives opacity and whiteness as well as controlling the flow of ink.

Single-crystal ceramics play important roles ranging from lasers to jewels. Several separate innovations have led to production of pure crystals in various shapes and sizes as well as the mass production at low cost of high-quality synthetic gemstones.

Glass–metal seals and glass metallization for other purposes became vital components of vacuum systems, including vacuum tubes for radio and other electronic circuits. One major innovation in this technology, glass–metal eutectic direct bonding, is presented here.

Ceramic coatings often are used to provide improved wear resistance. Flame spraying provides a means of applying heat-resistant and wear-resistant coatings to many substrates.

Medical and Dental Ceramics

Bioceramics

by James F. Shackelford, University of California at Davis

Bioceramics can be defined as engineered ceramics with applications in the field of medicine. We include in this definition some glasses and ceramic-based composites. Until the 1960s the inherent limitations of the mechanical properties of traditional structural ceramics limited their applications for medicine, despite that ceramics are more similar to natural skeletal materials than the more widely used metallic implants. Two of the more successful applications until that time had been dental porcelains and the use of tricalcium phosphate, a highly resorbable material, in conjunction with bone fracture repair. Improvements in ceramic processing technology in the 1960s led to a burst of interest in bioceramics. Sam Hulbert and co-workers were especially prolific, concentrating on the excellent biocompatibility of chemically inert oxide ceramics, such as alumina. In the early 1970s Larry Hench and co-workers developed Bioglass, a low-silicate glass composition with a phosphate addition helping to create a surface-reactive material and direct bonding to bone.

Despite promising research in the late 1960s and early 1970s, applications in the field progressed slowly. A less conservative governmental regulatory system in Europe led to substantially more clinical experi-

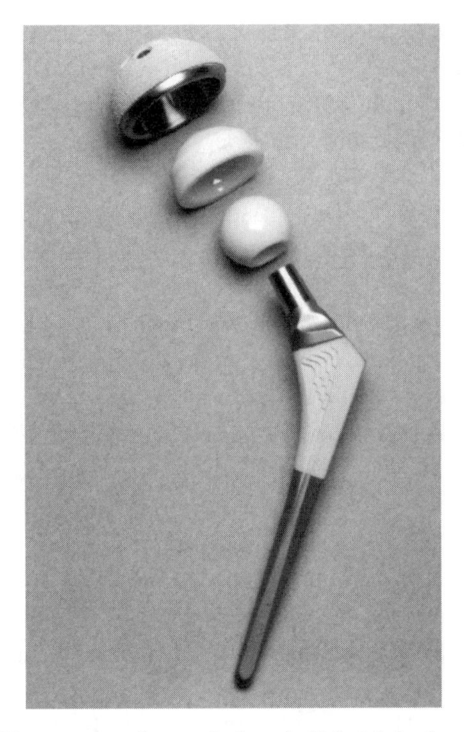

Figure 31. *Hip stem and acetabular shell (with hydroxylapatite coating) with alumina femoral bearing and acetabular insert. Courtesy of Osteonics.*

ence with bioceramics there. The past decade, however, has seen a renaissance of interest in bioceramics. This is well illustrated by the increasing use of plasma-sprayed hydroxyapatite coatings on metallic hip implants. The use of this ceramic, which is the primary mineral content of bone, represents the successful use of an engineered material in conjunction with its natural role in the body.

Another interesting and successful application of bioceramics is the repair of "bone defects," defined as centimeter-scale gaps in the skeletal system. The harvest of an autogenous bone graft carries significant morbidity and cost. Allografts (using cadaver bone) have problems with immunologic reaction and the risk of disease transmission. These limitations and concerns created substantial interest in the development of bone graft substitutes or extenders. An example is a composite composed of granules of a biphasic ceramic (hydroxyapatite plus tricalcium phosphate) in a matrix of collagen, the polymeric form of protein that constitutes about one-third of natural bone. An even more bonelike material is produced by the addition of bone marrow taken from the patient. Porous hydroxyapatite and granular tricalcium phosphate also have been used for this bone defect repair application. An active area of current research is the use of such bioceramics as delivery systems for bone morphogenetic proteins (BMP), shown to stimulate bone regeneration.

A wide variety of ceramics, glasses, and glass-ceramics have been used in dentistry, including the use of dental porcelains for almost two centuries. The largely compressive loads in dental applications have allowed the widespread use of Bioglass implants. Bioglass also has been used as a sound-transmitting prosthesis in ear surgery. Two separate bioceramic systems have proved useful for the treatment of cancerous tumors. Yttria aluminosilicate (YAS) glass beads have been used to provide localized radiation therapy in the liver. Alternatively, a ferromagnetic ceramic implant allows the use of an alternating magnetic field to provide localized heating of bone tumors, thereby killing cancer cells. An interesting area of recent research is the development of biomimetic materials that are produced in ways that imitate natural, biological processes. These low-temperature processes may be especially appropriate for biomedical applications.

Bioactive Glasses, Ceramics, and Glass-Ceramics

by Larry L. Hench, University of London

Replacing worn out body parts with bioactive glasses or ceramics that develop a high-strength interfacial bond to living tissues is one of the most innovative developments of the past century. Since the early 1960s, high-strength, corrosion-resistant metal alloys, polymers, and alumina ceramics have been used as implants to repair or replace damaged or diseased bones, joints, and teeth, thereby eliminating pain and improving the quality of life for millions of people. A common characteristic of these nearly biologically inert materials is that they become isolated from their host tissues by a thin nonadherant fibrous capsule that can eventually limit the lifetime of the device because of micromotion, loosening, and even fracture of the implant or the bone. In 1969 a special compositional range of soda–lime–phosphate–silica glasses was discovered that develop a mechanically strong, chemically bonded interface with living bone. These materials are called bioactive glasses. Only compositions in the center of the Na_2O–CaO–SiO_2 ternary diagram (region A with a constant 6 wt% P_2O_5, Fig. 33), bond to living bone.

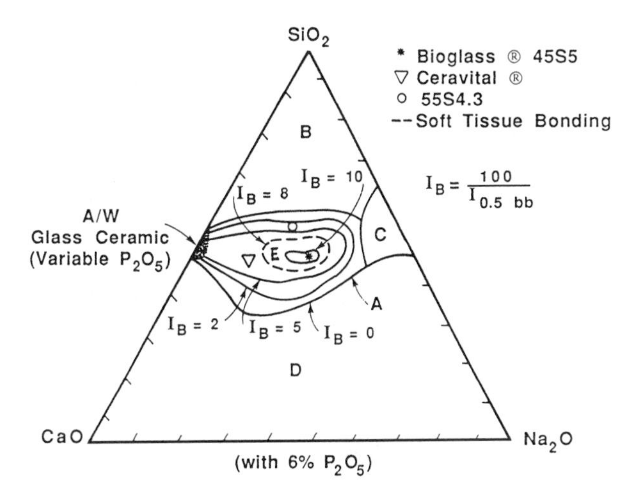

Figure 32. *Phase diagram of silica–lime–soda system. Courtesy of Larry Hench.*

Compositions in region C are too reactive and dissolve or resorb when implanted in bone. Compositons in region D are not technically practical. Compositions in the narrow range from 45% to 52% SiO_2, region E, bond rapidly to bone and also bond to soft connective tissues. Relative rates of bonding to bone are indicated as isobioactivity contours in Fig. 32. Compositions in the center with a high bioactivity index bond very rapidly, within days. Compositions near the bone-bonding boundary have a low bioactivity index. Bonding to living tissue occurs as a result of rapid ion exchange of alkali ions in the glass with hydrogen ions in body fluids leading to the creation of a hydrated silica gel on the glass and subsequent heterogeneous nucleation and crystal growth of a layer of biologically active hydroxycarbonate apatite (HCA) bone mineral on the implant surface. Cells attach to the surface layers, and collagen fibrils produced by the cells bond within the HCA crystal agglomerates and form a mechanically strong interface between living and nonliving materials.

Bioactive glasses are injection molded to make prostheses to replace the bones of the middle ear, or in the shape of cones to replace the roots of teeth and preserve the jawbone. Powders of bioactive glasses in the

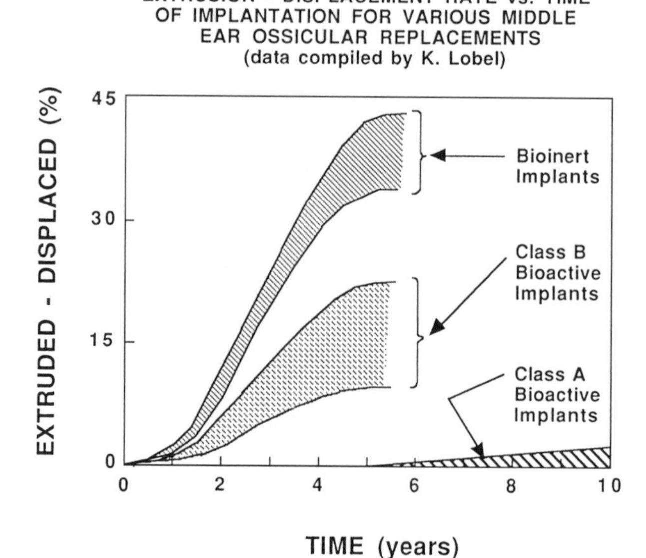

Figure 33. *Extrusion–displacement rate as a function of time of implantation for various middle ear ossicular replacements. Courtesy of Larry Hench.*

size range of 90–710 μm are placed in bony defects created by periodontal disease. The bone rapidly regenerates throughout the defects in a process termed osteoproduction and prevents the loss of teeth. Bioactive glass powders also are used to fill large bone defects following extraction of impacted third molars, to augment the jawbone prior to placement of dental implants or to regenerate new bone to stabilize total hip or knee implants, requiring revision surgery. High-strength, fracture-tough bioactive glass-ceramics made by sintering to high density apatite and wollastonaite phases bonded by a bioactive glass phase, called A/W glass-ceramics, are used clinically to replace vertebrae removed because of tumors, as vertebral prostheses, and as bioactive replacements for bone removed from the illiac crest for use as bond grafts. Bioactive ceramics in the form of plasma spray coatings of synthetic hydroxyapatite (HA) on metal substrates are used to enhance the bonding of bone to joint prostheses and dental implants. Synthetic HA of bioactive glass powders are used as a second phase in a polymer matrix to make bioactive composites with mechanical properties closely matching those of bone. Bioactive composites are used to replace middle ear bones and repair fractures of the skull. The advantage of all of these types of bioactive materials is that they bond to living tissues, and the mechanically strong interface stabilizes the "spare part" in the body. The important result of use of bioactive implants is that clinical survivability is much longer than previously used bioinert metal and polymer implants, as illustrated in Fig. 33.

Hydroxyapatite Coatings Assist Bone Bonding

by Karlis A. Gross, University of Sydney, and Monash University

Hydroxyapatite, a natural calcium phosphate biomineral found in teeth and bone, has been produced synthetically to provide a significant improvement in the performance of dental and orthopaedic implants. Traditionally, fixation of the implants was attained by use of mechanical fixtures, such as metallic screws, and later by using a grooved surface or a bone cement. Application of a thin layer of hydroxyapatite on the surface of implants by plasma spraying has provided a bioactive surface for direct bone attachment and, hence, a better fixation. Plasma spraying is the preferred coating technology, because it has fast deposition rates and is capable of producing a high coating density.

Hydroxyapatite was initially used as a powder or bulk implant in dental applications. In 1971 Monroe *et al.* noted the similarity between sintered hydroxyapatite and the mineral compositions of hard tissues and suggested it as a substitute for those tissues. The first U.S. clinical trials utilizing hydroxyapatite began in 1977. The ability of the hydroxyapatite to produce a strong bond with bone led to coating titanium alloy implants with this material. Clinical trials of hydroxyapatite coatings on femoral stems in humans were initiated by Furlong and Osborn in 1985 and Geesink in 1986.

Success of plasma-sprayed hydroxyapatite has been linked to its intrinsic biocompatibility and stimulatory effect on bone regeneration. Surface dissolution of the coating produces a saturation of calcium and phosphate species in the physiological media and is responsible for the osteoconductive effect that promotes rapid bone growth directed toward the implant. A higher bone coverage is attained compared to metallic implants of the same design. Bonding can be further enhanced with a coating on a porous surface into which bone can grow or with bone morphogenic proteins that enhance more bone growth. The coating is mainly used in applications such as dental implants, as well as knee and hip prostheses on threaded, grooved, and textured surfaces. Therefore, it can be used in conjunction with mechanical fixation mechanisms. This is especially important to increase the long-term stability in orthopaedic implants where the prostheses are weight bearing in contrast to earlier tooth implants. Hydroxyapatite coatings in general have facilitated the use in younger recipients who have a more active lifestyle and require a higher performance implant.

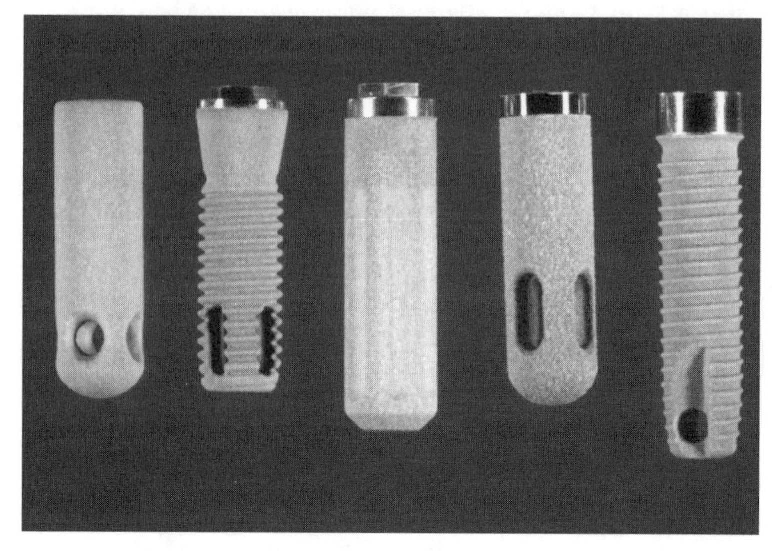

Figure 34. *Coated dental implants. Courtesy of Misch Implant Dentristry.*

Dental Ceramics

by J. Robert Kelly, National Institute of Standards and Technology and Dental Corps, U.S. Navy

Highly esthetic, durable, and biocompatible ceramics are available to replace missing teeth, to restore tooth structure lost to disease or trauma, and to substitute for discolored but otherwise healthy tooth enamel. Dentists and ceramic engineers began collaborations circa 1774, a time when only wood, animal products, and "donor" human teeth were available for tooth replacement. Ceramic dentures and denture teeth, although highly praised by medical societies of the late 1700s and early 1800s for their sanitary value, remained as fragile and unsightly as the whiteware and Parian formulations from which they were derived. Relatively lifelike denture teeth became available by the early 1900s as did polymeric denture bases that solved many processing and fit problems for removable prostheses. However, the use of ceramics for the repair of single teeth or in the replacement of a few teeth by a fixed (cemented) prosthesis was quite limited. Truly widespread application of ceramics in dental practice (along with major esthetic improvements) awaited two important structural innovations that did not occur until the latter part of this century. The first innovation created today's most often prescribed dental devices, metal–ceramic prostheses (crowns and bridges), enabling dentists to address simultaneously both structural and esthetic problems. Metal–ceramic prostheses are composed of alkali-modified aluminosilicate glasses (feldspathic) layered on thin (0.5–0.75 mm) castings of palladium-, gold-, or nickel-based alloys. Metal substructures proved crucial to achieving long-term clinical success, especially when ceramics were used to repair posterior teeth or to replace missing teeth with cemented prostheses.

Prior to the 1960s feldspathic dental porcelains were not thermally compatible with dental casting alloys; i.e., they would crack and debond from the metal during cooling. Weinstein, Katz, and Weinstein provided a key development in 1962 allowing the birth of this important technology, taking advantage of earlier observations by Orlowski and Koenig (1941, 1944) at The Ohio State University Engineering Experiment Station regarding the nonlinear thermal expansion of high-potassium feldspar glasses. Metal–ceramic compatibility was achieved via inclusion of crystalline leucite, a mineral having both high expansion/contraction behavior as well as a volume increasing/decreasing phase change occurring within the appropriate temperature range. In this manner the shrinkage during cooling was "matched" between the metal and ceramic, minimizing the thermal incompatibility stresses that cause cracking and debonding. In addition to making possible the widespread use of ceramics in dental practice, metal–ceramic technology fostered the development of a class of artisans skilled at replicating nature in dental ceramics.

The second structural innovation came in 1965 when McLean and Hughs extended dispersion strengthening to dental ceramics, creating the first successful substitute for metal castings. McLean and Hughs utilized alumina as the dispersed phase in a concentration and particle-size range that maximized strength and minimized opacity. Their contribution occurred at a time when basic principles of dispersion strengthening in glasses were just being appreciated from the literature, further demonstrating the ties binding dental materials and ceramic engineering scientists. McLean went on to clearly demonstrate the outstanding esthetics of all-ceramic prostheses compared to metal–ceramic prostheses and may be credited with creating a true renaissance among artisans, dentists, and scientists regarding the esthetic potential and further strengthening of dental ceramics. Esthetic improvements followed as ceramic usage became ubiquitous, leading to materials today that mimic the color, the translucency, and even the subtle opalescence of natural tooth structure. Universal recognition of the durable beauty and biocompatibility of ceramics has led in turn to recent processing research providing for the net-shape fabrication of esthetic ceramics, CAD/CAM, and precision milling systems that eliminate dental laboratory fabrication steps, and increasingly sophisticated routes to the substitution of metal substructures by polycrystalline ceramics.

Dental Restoration— Porcelain-Fused-to-Metal

by Paul J. Cascone, The Argen Corporation

One of the most important ceramic innovations of the past 100 years is most probably in your mouth. In 1959 a patent was granted to a Dr. Weinstein for a porcelain material that could be fired onto a metal alloy. Thus was born the porcelain-fused-to-metal dental restoration. This process was commercialized in the 1960s to a very eager market worldwide. Within a few years this process was the one of choice for dentists because it combined the look of natural teeth with the required strength of an alloy. This technique replaced the all-gold restoration and opened the way for esthetic dentistry. Today in the United States, Canada, and Europe >80% of all dental prosthetic restorations are fabricated using the porcelain-fused-to-metal technique.

Over the years, the porcelains have evolved into more natural looking materials having improved translucency, shading, and florescence. The strength of the porcelains also has been the subject of development efforts. Soft porcelains have been introduced that do not erode the opposing natural dentition (compared to their predecessors). Some materials have been developed to a point were no alloy is needed as a substructure. These all-ceramic materials are pressed into shape and are used mostly for replacing single teeth.

Sol–Gel Transplantation Therapies

by Edward J. A. Pope, Solgene Therapeutics, LLC

Solgene Therapeutics, LLC, has developed a new biotechnology drug delivery system through the application of advanced sol–gel-processing technology that enables the encapsulation and transplantation of live therapeutic protein secreting tissue cells without the need for immunosuppression. Unlike conventional alginate approaches, sol–gel silica microbeads are porous enough to allow the host (patient) and implanted cells to exchange nutrients, waste products, and secreted therapeutic proteins, but exclude lymphocytes and antibodies that the host's immune system uses to attack foreign cells. In initial research in the area of diabetes, encapsulated, healthy pancreatic islets were transplanted into diabetic mice, resulting in the elimination of all hematological and urological manifestations of the disease for more than four months. Further research on genetically engineered cells is in progress.

Through the encapsulation and transplantation of genetically engineered cells, the biotech "factory" can be transplanted into the patient, thereby manufacturing the drug *in vivo* for the duration of therapy. The duration of therapy can be regulated by the timed breakdown of the encapsulating agent thereby exposing the transplanted cells to the immune system, by the preprogrammed death of the cell (apoptosis), and by external chemical triggers to cell death. Thus, biodegradable silica-gel encapsulation represents a potential drug delivery platform for delivering hormones, biotech drugs, insulin, and detoxifying agents. It also can be used for livestock and animal health applications.

The chemical routes to processing silica-gel at temperatures near ambient by the sol–gel process is well understood. Hydrolysis of tetraalkoxysilanes occurs most rapidly at low pH, whereas polycondensation is most rapid at high pH. The gels form by a fractal process of cluster growth followed by cluster–cluster aggregation. In order to encapsulate living cells, a substantially modified method of Brinker was used. The microstructure of the resulting gels reveals an aggregate of ~8.5 nm clusters with pores of approximately the same order of magnitude. The preferred geometry for cell encapsulation is small microspheres. Microspheres have been made by both emulsion and by drop tower in a biologically nontoxic oil for which the gel-forming solution is immiscible. Both emulsion- and drop-tower-derived silica-gel show average pore diameters (in the wet hydrogel state) of between 10.5 and 16.1 nm. Known cell pathogenic agents, from antibodies to protozoa, range in size between 50.0 nm to >10.0 μm.

In encapsulating living cells in silica-gel, several key factors need to be addressed, including pH, temperature, salinity, and atmosphere (aerobic versus anaerobic). Our earliest attempts at cell encapsulation involved the common fungi S. cerevisiae, in which spores of the fungi were encapsulated in silica-gel and showed bioactivity even after a one year hold at 5°C. Dubbed by *Business Week* "A Ceramic Biosphere for Benign Bugs," our early research into fungi pointed the way to ever more complex and challenging research in drug delivery and diabetes. Since then, numerous fungi and bacteria as well as a wide range of mammalian tissue have been successfully encapsulated, showing long-term viability and potential therapeutic value.

In order to investigate whether silica-gel encapsulation methods can provide an attractive alternative to the more conventional alginate materials that currently predominate, an *in vitro* and *in vivo* experimental protocol was established. Many of the specific details have been published previously. Donor tissue was harvested from C57ob/ob mice, followed by enzymatic digestion in collagenase/dispase solution. Upon verification of islet viability, islet suspensions were encapsulated as described previously and cultured in growth media (bovine growth sera) and subjected to further viability tests (FSR). The prefered capsule size is <500 μm to permit ample diffusion of metabolites to all beta cells in the capsules.

Viable capsules were transplanted into male NOD mice that were rendered diabetic through the administration of streptozotocin. All transplant recipients had a blood glucose level >400 mg/dL prior to implantation. Our earliest success was dubbed "supermouse" with a complete elimination of measured hyperglycemia for more than two months. It was determined that the duration of treatment was limited by the time required for the capsule to biodegrade. Subsequent tests with an improved silica-gel formulation resulted in similar success in excess of 4.5 months in a study involving more than 30 diabetic mice!

A new and promising delivery platform, based upon sol–gel silica encapsulation of living cells, has been demonstrated. Pore size is sufficient to simultaneously allow large biomolecules to pass while excluding the larger antibodies of the immune system. Preliminary diabetes studies demonstrate the potential for long-term therapy for diabetes. Genetically engineered cells designed to secrete biomolecules also have been demonstrated. Current research on bone regeneration, involving a 95 rat study at Cornell, hold great promise. This technology offers competitive advantages for the delivery of varied biotech drugs.

Ceramic Scintillators for Medical X-ray Detectors in Computed-Tomography Body Scanners

by Charles D. Greskovich, General Electric Company

An X-ray computed-tomography (CT) scan is now a common diagnostic procedure in hospitals and clinical sites to image selected regions inside the human body for detection of cancer and other diseases. A typical CT scan takes only about one second, and during that time an X-ray detector in the system collects almost one million readings of the radiation passing through the patient's body. These signals are fed to a computer and translated into a cross-sectional image of the body for display on a video monitor. The X-ray detector is a very important component that must have high efficiency to provide low-contrast detectability and low-noise images. In 1988 General Electric Medical Systems (GEMS) unveiled an ultra-high-performance detector for its top-of-the-line CT scanners. A dramatic improvement in the image quality of organs and structures in the human body was made possible by the new X-ray detector. The key to the detector breakthrough was a new ceramic scintillator material developed specifically for this application by scientists at the GE Research and Development Center in Schenectady, NY. It took more than four years to invent the preferred composition and ceramic processing procedures for the ceramic scintillator before transitioning the overall processes to GEMS for scale-up and manufacturing into efficient solid-state X-ray detectors. A team of physicists, materials (ceramic) engineers, chemists, mechanical and electrical engineers, and computer programmers had to work closely together to make the advanced detector and CT scanner a reality.

The detector is constructed with almost one thousand bar-shaped scintillator elements, ~30 mm long, 1 mm wide, and 3 mm deep, each coupled to a photodiode. The approximate composition of the ceramic scintillator is $Y_{1.34}Gd_{0.60}Eu_{0.06}O_3$ and is comprised of rare-earth oxides, each imparting special properties. When struck by medical X-rays, the ceramic elements glow, or scintillate, with red light emitted by the european ions in the crystalline ceramic. This light is sensed by the photodiodes that feed electrical signals to the computer for image reconstruction. GE's ceramic scintillator is prepared from fine, chemically uniform powder of purity >99.99% that is formed into powder compacts. A special high-temperature sintering process was developed that transforms the porous compacts to fully dense, optically transparent ceramics. The transparent material has a polycrystalline microstructure, which permits precise and reproducible machining of the scintillator elements to correct dimensions, something that is very difficult to do with the competitive single crystals of CsI and $CdWO_4$ scintillators used by other CT system manufacturers. The ceramic route to fabricating transparent scintillators offers great flexibility by use of high-purity chemical components and special additives to alter and control some of the critical properties, such as scintillation efficiency, optical transparency, X-ray damage resistance, and X-ray absorption. To this day more than five thousand CT scanners with X-ray detectors containing ceramic scintillator/photodiode arrays have been sold worldwide. This efficient ceramic scintillator has generated a significant advancement in CT medical imaging and has allowed both the business to gain sales growth and the detection of cancer at earlier stages without the risk of increased X-ray dose.

Nuclear and Environmental Ceramics

Uranium Dioxide (UO_2) as a Nuclear Fuel

by Ian J. Hastings, Atomic Energy of Canada, Ltd.

Prior to 1940 the commercial applications of UO_2 comprised limited use as a coloring agent in glass and ceramic glazes. In 1941 seven tons of compacted UO_2 powder was used in the exponential lattice experiment at Columbia. Forty tons of pressed natural UO_2 was used as a major part of the Stagg Field self-sustaining chain reaction in December 1942. The main argument for continuing development of UO_2 as a nuclear fuel was that it possessed a simple cubic fluorite lattice and thus would be less prone to some of the difficulties being experienced at the time with metallic uranium—irradiation growth, thermal-cycling effects, and fission-gas swelling. UO_2 also was attractive because of its relative chemical inertness in contact with coolants, such as water.

Undoubtedly, the major impetus for UO_2 as a nuclear fuel, and one that ultimately led to its almost exclusive use in commercial reactors, was the instigation in 1953 of the Pressurized Water Reactor (PWR) Project under Admiral Rickover. The then Atomic Energy Commission and the Joint Congressional Committee on Atomic Energy decided on the construction of a pressurized water reactor to be in operation by the time of the Second International Conference on the Peaceful Uses of Atomic Energy in Geneva in 1958. This was a commercial follow-up to the Mark 1 prototype submarine reactor plant in Idaho, and the *Nautilus* submarine propulsion plant.

The Bettis Laboratory, operated by Westinghouse, was chosen for the PWR plant design, development, and construction. Of a number of fuel candidates, UO_2 in the form of Zircaloy-clad, cold-pressed, and sintered pellets, was ultimately the successful one. The development program examined the production and characterization of sinterable UO_2 powder, pressing, and sintering to produce pellets with densities in excess of 92% of theoretical, and irradiation testing. Current UO_2 fuel, with some improvements, is basically the same as that original product. The Shippingport core was completed, shipped, installed, and began supplying electricity to the Duquesne Light Company system December 18, 1957.

It is important to acknowledge the critical contribution of the international collaboration in proving UO_2 as a practical nuclear fuel between the concurrent U.S., U.K. (advanced gas-cooled reactor), and Canadian (CANDU® pressurized heavy-water reactor) programs, and the important role played by the irradiation test loops in the NRX reactor at Chalk River in Canada. At least three current and former American Ceramic Society members played pioneering roles in the development of UO_2: Jack Belle, David Kingery, and Ross MacEwan.

In November 1997 there were 444 nuclear power plants operating worldwide (supplying 17% of the total electricity production), and 36 under construction, with 110 in the United States (22% of U.S. electricity production). The overwhelming majority of these plants use natural or enriched UO_2 as fuel. Other countries have an even larger investment in nuclear electricity: France 77%, Lithuania 83%, Belgium 57%, Slovak Republic 45%, and Ukraine 44%. The net worldwide electricity capacity based on nuclear is 351 GWe. Nuclear continues to be an important component of current and future environmental sustainability: in one year, a 1000 MWe coal plant produces 6.5 million tonnes of CO_2, more than 10,000 tonnes of other greenhouse gases and particulates, and 1 million tonnes of ash containing 400 tonnes of heavy metals. In the same time, an identical nuclear plant produces 35–100 tonnes of spent fuel and zero other emissions. The spent fuel can be safely and securely stored until disposal methodologies are finalized.

Ceramic Fuel for Space Exploration

by D. Thomas Rankin, Westinghouse Savannah River Company

A special isotope of plutonium, ^{238}Pu, is currently supplying power for the Galileo spacecraft now orbiting Jupiter and exploring this planet and its four moons. Likewise, the Cassini spacecraft is on its way to Saturn on a similar mission. Long-term, deep-space missions, such as Galileo and Cassini, require a nuclear source to provide the electrical power needed to operate the instruments on board the spacecraft. Solar cells do not work because the distance from our sun is too great, and the corresponding solar energy per unit area is too low. Chemical batteries have unfavorable weight and lifetime issues.

The United States has used and continues to use successfully radioisotopic thermolectric generators (RTGs) to supply electrical power for deep-space missions. The RTGs consist of a nuclear heat source and a converter to transform the heat energy from radioactive decay into electrical power. ^{238}Pu serves well as the heat source because it produces 0.5 watts (thermal) per gram by emitting α radiation, and its half-life of 87.4 years provides for reliable and relatively uniform power over the lifetime of the mission. The RTG also is a very reliable power source because it has no moving parts.

The preferred chemical form for the heat source is $^{238}PuO_2$, a ceramic. This face-centered cubic oxide offers excellent chemical stability, a high melting point (>2450°C) and chemical compatibility with its container material, an iridium alloy. The absence of phase changes in this

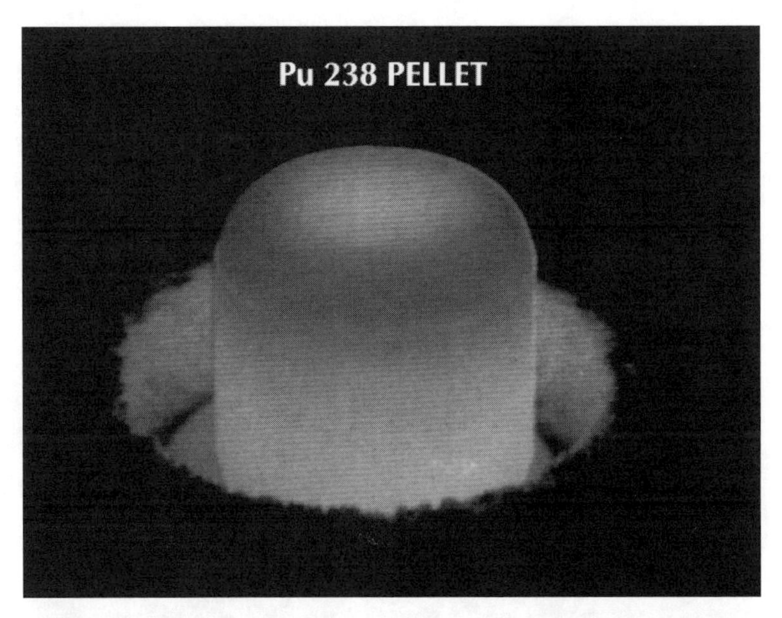

Figure 35. *^{238}Pu pellet. Courtesy of Westinghouse Savannah River Company.*

material facilitates fabrication and enhances the integrity of the heat source during processing and its eventual use at an operating temperature of ~1350°C. A fabrication process, developed at the Savannah River Site, produced 150 g cylindrical pellets, ~2.7 cm × 2.7 cm, each of which generates 62.5 watts (thermal) at the time of production. Each RTG produces 285 watts electrical converted from the heat decay of 72 of these $^{238}PuO_2$ pellets. The RTGs also are compact (0.43 m in diameter × 1.13 m in length) and low in weight (55.5 kg). The two RTGs on board the Galileo spacecraft supply all of the electrical power required by the spacecraft.

The fabrication process consists of hot-pressing a blended mixture of sintered $^{238}PuO_2$ granules prepared from calcined plutonium oxalate powder. Hot pressing provided the dimensional control needed, as the nominal pellet density was 84.5%. Both the granule-sintering and hot-pressing conditions required close control to minimize cracking and to permit ejection of the pellet from the hot-press die. A uniform density and the distribution of large intergranule porosity in the microstructure provide for the dimensional stability and the release of decay helium required at the elevated use temperatures.

NASA recently extended the very successful Galileo mission from two to four years. Many photographs are being taken and scientific data are being collected by the instruments of board Galileo and feedback from the probe that parachuted into the Jovian atmosphere. The exploration of Saturn by Cassini comes next. All on-board electrical power to operate the cameras, collect the data, and relay information to Earth originates from the ceramic fuel.

Figure 36. *Galileo as launched from space shuttle. Courtesy of NASA.*

Nuclear Waste Glasses

by George G. Wicks, Westinghouse Savannah River Company

Glass is the material of choice of every major country in the world today for immobilization of nuclear high-level waste (HLW). Nuclear waste glasses represent a special class of ceramics that possess many characteristics important to safe and effective long-term management of radioactive materials. These glass systems can be tailored to incorporate, and ultimately isolate, a wide range of nuclear as well as nonnuclear hazardous elements and, hence, protect the public and environment from the potentially detrimental effects of nuclear waste and other types of hazardous materials.

There are a variety of factors that contribute to the suitability of glasses for immobilization of HLW. High-level radioactive wastes can be generated from reprocessing of spent nuclear fuel used to generate electricity (commercial HLW) as performed overseas, or produced from the production of special nuclear materials used for national defense (defense HLW), as conducted over the years in the United States. In general, factors contributing to the suitability of immobilizing HLW into glass fall into two major categories. The first category involves good technical performance in five major areas of interest: flexibility/waste compatibility; mechanical integrity; thermal stability; radiation effects; and chemical durability. Chemical durability is generally considered the most important technical property of waste glass and provides a measure of how well radionuclides are tied up in the random network structure of the glass, even when subjected to possible leachants, such as groundwaters in a repository setting. The second important category is processibility, which involves the ease of being able to produce waste forms even under difficult remote conditions, necessary for processing highly radioactive material. The only HLW immobilization facilities in production in the world today are glass-making operations. These include the successful facilities at Marcoule and LaHague, France, a plant in Mol, Belgium, and the vitrification facility in Sellafield, England. The first major vitrification facility was the French plant at Marcoule, which began operation in 1978. There also are additional waste vitrification plants either in production, under construction, or being planned in other countries, including Japan, the former Soviet Union, Germany, India, and the United States.

The United States has ~100 million gallons of HLW containing well over 1 billion curies of radioactivity. This waste has resulted primarily from defense operations and activities since the Manhattan Project of World War II. In April 1996 the more than one billion dollar Defense Waste Processing Facility, or DWPF, began radioactive operations at the Savannah River Site in Aiken, SC, and has produced more than 1,000,000 pounds of nuclear waste glass product thus far. The ability of borosilicate glass to incorporate and isolate nuclear waste and then to be disposed of, as part of a multibarrier isolation system in carefully selected geologic repositories, provides the final steps in closing the nuclear fuel cycle. This disposal system provides the necessary isolation of potentially hazardous material for our generation and for all generations to follow.

Honeycomb Ceramics

by James R. Johnson, 3M Company, retired

A major development in the control of air quality, worldwide, is the use of catalytic converters in automobiles and similar devices in industry. The most widely used converter in automobiles depends on a honeycomb ceramic support for the catalyst. This device was invented in the late 1950s by J. R. Johnson, T. S. Reid, and D. O'Brien of 3M Company, using honeycomb ceramic technology they had developed based on tape-cast ceramic sheets that were corrugated, assembled with slip as a glue, and sintered into a monolith. These sheets evolved from the tape process invented by J. Parks and others at American Lava/3M. A similar honeycomb technology was developed independently at almost the same time at Corning Incorporated by R. Hollenbach, using a process whereby tea-bag paper was soaked in ceramic slip, corrugated, and made into monoliths. Many experimental products were developed in the 1960s with these new materials, but the catalytic converter had to wait more than a decade until a law was passed taking lead out of gasoline. Lead formed a glassy coating on the surface of the catalyst and destroyed its function. Englehard Corporation was very helpful in the development and application of suitable catalysts that included incorporating a coated alumina substrate on which was deposited platinum or platinum-group elements. 3M Company made the ceramics for the first honeycomb converters used in fork lift trucks and for the 1974–1975 automobiles using sheets of corrugated cordierite ceramic made by a novel ceramic paper process developed by R. A. Hatch. Corning and 3M both developed an extrusion process to make honeycomb, but because 3M was divesting itself of much of its ceramic businesses (American Lava Corporation) at that time, it discontinued its converter business. Corning became the major producer, as it is today, of converters using their extrusion process, developed by R. Bagley. The extrusion dies and process were intricate and complex in comparison with standard extrusion technology of the time.

There were many other applications envisioned for the honeycomb ceramics. At 3M the initial driving force was the desire to make high-temperature nuclear fuel elements. Heat exchangers, catalyst supports, filters, and the extension of the technology to metals and composites soon followed. Corning Glass produced some excellent rotary heat exchangers for experimental engines in the 1960s using the tea-bag-paper process. There have been many other applications since, actual or envisioned, and the uses of high-surface-area ceramics in honeycomb and other forms (beads, pellets, fibers, coatings) are a major and growing part of the technical ceramics family.

Honeycomb Ceramics

by Margaret K. Faber, Corning Incorporated

The largest single use of honeycomb ceramics today is as supports for catalysts for emission control devices in automobiles, for both diesel and gasoline engines. This important industrial application of ceramics was first developed in response to the Clean Air Act of 1970. In the 1960s air quality had deteriorated significantly from acceptable standards, and more than 50% of the carbon monoxide and hydrocarbon pollutants present in the ambient air was generated by automobiles. The Clean Air Act required that all new automobiles had to meet strict emissions standards for carbon monoxide and hydrocarbons by 1975, and additional emission standards for nitrogen oxides by 1976.

Several innovations led to the success of the honeycomb catalytic converter product. Among these was the development of lead-free gasoline, which enabled the use of precious-metal catalysts to oxidize the carbon monoxide and hydrocarbons to carbon dioxide and water. Moreover, Corning Incorporated (then Corning Glass Works) and the American Lava/3M Company had independently invented ceramic monolith structures that were made by wrapping tape-cast ceramic sheets into a cellular structure akin to rolled up sheets of corrugated cardboard. These presented an intriguing alternative to packed beds of ceramic beads as supports for the precious metal catalysts. The primary advantage was the reduction of the pressure drop as the exhaust gases passed through the honeycomb converter rather than a packed bed. This ensured that the engine performance would not be compromised.

Because the honeycomb structures needed to withstand the rugged environment of the automobile exhaust system, the technical requirements were challenging. Corning Incorporated began an intensive research, development, and engineering effort to develop a viable ceramic catalytic converter using a relatively-low-cost process. In particular, the ceramic materials had to be refractory, strong, and able to withstand large and rapid variations in temperature without cracking. They had to be porous so that the catalytic coatings could be readily applied. Finally, these ceramic honeycombs had to provide a low pressure drop and a high geometric surface area for the catalyst coatings.

R. D. Bagley of Corning Incorporated made an important invention for a novel die structure for extrusion of thin-wall cellular structures with openings or passages that extended longitudinally through the body. This meant that the honeycomb supports could be manufactured readily, and with a variety of outer dimensions. Moreover, the individual cells were very open thus allowing a sufficient catalyst loading to be applied by a coating process, while maintaining low back pressure in the exhaust system. To address the ceramic materials issues, another important invention was made by I. M. Lachman and R. M. Lewis. Their invention was an extruded honeycombed cordierite material whose microstructure was characterized by oriented crystallites resulting from the orientation of the precursor clay material used to extrude the body. This anisotropic sintered ceramic material had a low coefficient of thermal expansion, resulting in a very high resistance to thermal shock. The cordierite honeycombs also were strong, lightweight, and could be made with controlled porosity.

Corning's early Celcor products had 200 cells/in.^{2} with a wall thickness of ~10 μin. Today, the need for lower back pressure and faster rates to reach catalytic conversion temperatures has led to the development of higher cell density and thinner wall products. Products with 400 cells/in.^{2} and 6 μin. walls, and 600 cells/in.^{2} and 4 μin. walls are used today. Applications of honeycomb ceramics include diesel particulate filters, wood-stove catalytic combustors, and molten-metal filters. Other materials such as honeycombs of molecular sieve zeolite structures have been formed either by direct extrusion, by coating the cordierite with a zeolite layer, or by directly growing the zeolite crystals in the walls of the honeycomb. New applications for honeycomb catalysis technology can be found outside the automotive industry. For example, honeycomb ceramics can be used in chemical processing applications where they may enable processing innovations that could lead to substantial energy savings or improvements in yield. Carbonaceous honeycombs also are being used to adsorb and filter pollutants from water sources.

Other Special Ceramics and Techniques

Inorganic Ceramic Membranes

by Anthonie J. Burggraaf, University of Twente, retired

Since the late 1980s R&D in inorganic membranes has increased strongly because of their special properties in separation processes and chemical reactors. Different types of dense (nonporous) and porous inorganic membranes exist. By far the most important subgroup are asymmetric, composite ceramic membranes in tubular, disk-shaped, or stacked plate modular systems.

The membrane system consists of several components. The mechanical stability is achieved by a 1–2 mm thick (meso) porous support with a pore diameter d_p in the range 1–15 μm. On top of it may be deposited one to three thin (mesoporous) oxide layers with decreasing thickness of tens to a few micrometers and decreasing pore size from 1000 to a few nanometers. The mesoporous top layer can be coated with a very thin microporous layer (d_p of 0.4–0.8 nm), or it can be loaded with a catalyst.

Alternatively, the top layer can consist of a nonporous oxide or metal layer, semipermeable for hydrogen and/or oxygen. The top layer always determines the required high separation efficiency for separation of liquid or gas mixtures.

Crucial are the pore size, which should be as uniform as possible, and the absence of defects in the top layer. The top layer should be thin to promote the maximum obtainable flux. These complex architectures are necessary to combine the requirements of high separation factors and large fluxes and have been developed from the single macroporous support systems.

Innovative steps included the development and structural analysis of defect-free, thin (2–4 μm) mesoporous (d_p of 2–4 nm) γ-alumina top layers (1981) and of defect-free, ultra-thin (< 0.1 μm) microporous (d_p of 0.4–0.7 nm) silica/titania top layers for gas separation (1991), both by Burggraaf and co-workers. Some years later zeolite top layers were simultaneously developed by several groups in Europe, the United States, and Japan.

Industrial production of ceramic membrane systems for microfiltration of liquids was first reported by a French industrial group (Ceraver) in 1984. Liquid filtration forms the most important application field today and includes treatment of wastes (e.g., oil emulsions and bilge water), recycling/reuse of materials (e.g., solids, paint/ink, and broths from biotechological processes), and processing (e.g., fruit juices, wine clarification, and whey). Ceramic membranes also find application in an ever-increasing number of air-separation, gas-separation in petrochemical processes, and syn-gas synthesis.

Ceramic Water Filter

by Robert Roth, Marathon Ceramics

Safe water is a worldwide concern. Almost 20,000 children in third-world countries die each day from unsafe water, and outbreaks of cryptosporidiosis in U.S. city water supplies have resulted in more than 100 deaths in Milwaukee and dozens in Las Vegas. For campers in remote locations of the United States, it is wise to process water before drinking it. An inexpensive and effective ceramic water filter can significantly improve the health of millions of people worldwide, both at home and while traveling.

Mountain Safety Research set out to design a water filter appliance in 1991 that could be used for backpackers and that would supply a low-cost and effective filter solution to customers worldwide. For more than 130 years, ceramics have been used for water filtration using, fundamentally, the same manufacturing technology. But the products available in 1991 were either too expensive or too low in quality. One of the key concerns was maintaining tight control over dimensional tolerances.

After two years of research, Dan Vorhis and Chris Adams of Mountain Safety Research, working with Professor O. J. Whittemore of the University of Washington, developed a new ceramic water filter manufacturing process using extrusion. The result is a low-cost and high-quality product that has been proved effective as a water filter under extreme conditions. The extrusion process is covered under U.S. Pat. No. 5,656,220 and is incorporated in the MSR MiniWorks™ and MiniWorks™ II water filters.

In 1997, the U.S. Marines selected the MSR MiniWork filter as standard issue for their Raids and Reconnaissance Division. After reviewing several water treatment options and analyzing microbiological tests, the Marines selected the MiniWorks product with the MSR ceramic filter technology and placed a major order for delivery in 1998.

From a recent E-mail message, one of the MSR customers wrote: "Dan, I just returned from Brazil where I spent six weeks in the field in which only the MSR was used for water filtration. It worked well overall. I have loaned one of the units to a student who is in Kenya, Africa, for six months. He reports after four months that he is the only one in his group that has remained healthy, but he is using the MSR daily while others have different filter units or use chemical treatment of the water." Wayne Carmichael, Professor Aquatic Biology/Toxicology.

For filtration of water biohazards, this new ceramic manufacturing process places the cost of an effective water filter within the reach of any mass consumer market.

Low-Expansion Ceramics

by Rustum Roy, The Pennsylvania State University

The first negative expansion crystalline ceramic phase that eventually became useful was found by F. A. Hummel in 1947 in a sample of fired petalite. Eventually, the active component was shown to be β-eucryptite and a patent was issued to Hummel and The Pennsylvania State University. The complete phase relations, including the subsolidus relations involving the natural materials were elucidated by Rustum Roy and E. F. Osborn, and the detailed thermal expansion characteristics were determined by Hummel and his students Bush and Gillery. This material eventually became the key component in Corning's pyroceram, and other glass ceramics.

A small list of ceramic phases have been found since that have a zero or nearly zero thermal expansion. Most of these are anisotropic crystals that expand slightly in one direction and contract more in the other because of appropriate twisting and displacement of the polyhedra. ZrW_2O_8, discovered 30 years ago, is a rare cubic example. However, most of these phases are not of interest because of their low temperatures of reaction and of melting.

In 1982 Roy, Alamo, and Agrawal reported the discovery of the largest structural family of materials with tailorable thermal expansion from modestly negative through zero to low positive values. This family was named NZP representing the parent compound ($NaZr_2P_3O_{12}$). The NZP structural family was found to consist of more than 150 members, many of which have been characterized by their ultralow thermal expansion behavior. This perhaps is a unique structural family in which by careful ionic substitution one can develop a composition of virtually any thermal expansion value over a wide temperature range suitable for a specific application. The NZP family permits the engineering of a variety of properties, such as candidates for devices requiring high thermal shock resistance, catalyst supports in the automobile industry, hosts for nuclear waste, oxidation protection coating material for carbon–carbon composites, and, recently, active catalyst for NO_x reduction.

Phase Diagrams for Ceramists

by Stephen Freiman, National Institute of Standards and Technology, and Rustum Roy, The Pennsylvania State University

The most generally used fundamental research in all ceramic science and technology is the phase equilibrium diagram describing the formation, coexistence, stability, and melting behavior of any given composition. Following J. W. Gibbs, this has been a strikingly American area of basic science. The lion's share of the credit belongs to the Geophysical Laboratory of the Carnegie Institution of Washington. It was there that L. H. Adams, N. L. Bowen, J. Greig, J. F. Schairer, J. W. Morey, H. S. Yoder, Robert Sosman, and many others did the meticulous physicochemical studies we call phase diagrams. A few city blocks away, at the (then) National Bureau of Standards (NBS), H. F. McMurdie and E. M. Levin started to collect and collate and make these diagrams available to the world. Moreover, a distinguished alumnus of the Geophysical Laboratory, E. F. Osborn, moved to The Pennsylvania State University and there started what became the Academic Center for Phase Diagram Determination. His students, Rustum and Della Roy, R. C. DeVries, A. Muan, and F. A. Hummel, created a formidable array of phase diagram determiners. Osborn and Muan pioneered in the multivalence state systems. Roy, Osborn, and Tuttle designed the key pressure vessels that made possible the determination of systems containing water and other volatiles. The records are plain for all to see in the index of names, in the work continuing at the NIST under the leadership of R. S. Roth in conjunction with The American Ceramic Society. The results are available as a series of books in the *Phase Diagrams for Ceramists* series and in computer format.

The first compilation of "evaluated" ceramic phase diagrams was published by F. P. Hall and H. Insley as the October issue of the *Journal of the American Ceramic Society* in 1933. Further compilations of phase diagrams were published as *Journal* issues or as companions in 1938, 1947, 1949, 1956, and 1959. The first volume in the currently (1998) 12 volume, hard-cover series *Phase Diagrams for Ceramists* was published by The American Ceramic Society in 1964, authored by E. M. Levin, H. F. McMurdie, and C. F. Robbins of NBS.

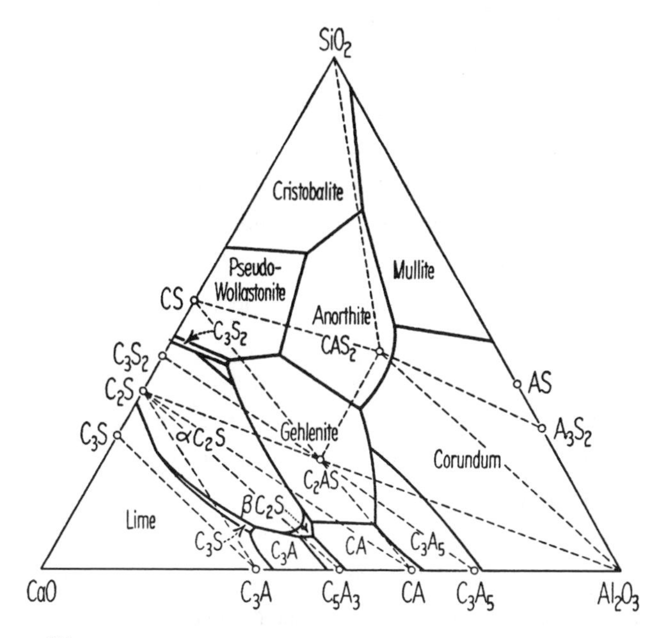

Figure 37. *Silica–calcia–alumina phase diagram. Courtesy of NIST and The American Ceramic Society.*

In 1982 a formal agreement was signed between The American Ceramic Society and NBS that outlined the duties of each party in the publication of evaluated phase diagrams. The American Ceramic Society, under the leadership first of James R. Johnson and later of R. M. Spriggs, undertook a campaign to raise funds to ensure the permanence of the ACerS/NBS Phase Diagram Program. Approximately $1.8 million was raised from companies, universities, individuals, etc., which supports The American Ceramic Society activities in the program.

The first computerized version of phase diagrams, offering search and graphic manipulation capability, was offered for sale in 1993. In 1995 all of the published diagrams were placed on a CD-ROM. To date, more than 53,000 copies of phase diagram volumes have been sold. A study being conducted by NIST shows that patent references suggest that the availability of the evaluated diagrams has made a significant impact on the development of new ceramic materials. All in all, this program represents a unique partnership between industry, academia, and government to provide critical data to the ceramics community.

Thermoluminescence Dating of Ceramics

by Christopher Maurer, State of Washington Department of Ecology

A major problem archaeologists have long faced is determining the age of the materials that they find. The development of carbon-14 dating after World War II helped, but this method could only be used for materials containing carbon. Archaeological ceramics, because they do not contain carbon, could only be dated by their shape and the style of their decorations. Such age estimates were subjective and were frequently disputed by other scholars.

In the 1960s building on the work of Farrington Daniels 10 years earlier, Martin Aitken, Stuart Fleming, and others extended the application of thermoluminescence measurement to the dating of archaeological ceramics. Thermoluminescence, (light given off by heating), refers to light emitted by mineral crystals, primarily feldspars and quartz, found in clays. Nuclear radiation from the soil surrounding the mineral crystals causes electrons in the crystals to become trapped, storing energy. When the crystals are heated to 500°C, the electrons are released, and the stored energy is emitted as light. The intensity of this light can be measured using a photomultiplier tube. The more nuclear radiation the crystals are exposed to, the greater the intensity of light emitted. Because the rate of nuclear radiation emitted by the soil is constant over long periods of time, a greater radiation exposure means a longer time of exposure and an older age for the archaeological ceramic. As long as the oven temperature exceeds 500°C, the firing of the ceramics at the time of their manufacture removes all of the energy stored in the mineral crystals up to that time. Thus, the age measured by thermoluminescence is the age since the ceramic was made and not the age since the mineral crystals were first formed. The technique requires measurement of the dose rate of ionizing radiation encountered by the specimen as well as calibration of the specimen in the laboratory.

The application of thermoluminescence is having an impact on archaeological dating. For the first time, it is possible to directly date pottery fragments, ceramic materials found at most archaeological sites, and, frequently, the only material found at such sites. The establishment of reliable dates for archaeological sites enables archaeologists to better understand which sites had developed first and which sites had developed later, thus showing how nations and ideas had spread or failed to spread beyond their origin.

The application of thermoluminescence dating also is important for museum conservation. The value of a museum object can vary by millions of dollars, depending on whether it is original or a modern imitation. Some ceramic wares, such as those from ancient China, are among those frequently imitated. Until the development of thermoluminescence dating, only expert opinion could determine if an object was genuine or not. The question for a museum frequently is whether an object is several thousand years old or less than a hundred years old, rather than a precise date, a task for which thermoluminescence dating is more than adequate. A critical advantage for the use of thermoluminescence dating by museums is that thermoluminescence dating can be done using only a few milligrams of clay. Such a small amount of material could be removed from the object without damaging its artistic integrity.

An example of such museum dating is that of T'ang figurines. These molded ceramic figures, usually coated with beautiful multicolored glazes, were originally made during the eighth century T'ang dynasty in China. They were used as funerary objects to be buried with the dead. During the first years of this century, the figurines began to be found and immediately became popular with art collectors and museums. Found at the same time were the original molds used to manufacture the figurines. As a result, the collector's market was flooded with modern imitations. Because the original molds were used, it was almost impossible to identify the imitation figurines from the genuine. Museums were left with uncertainty whether their collections were genuine. With the development of thermoluminescence dating, this uncertainty finally could be resolved, reassuring many museums that they did indeed have genuine T'ang figurines in their collection and enabling the modern imitations to be properly labeled as such.

Tubular-Sheathed Heaters

by Marcus Borom, General Electric Company

Since the 1920s tubular-sheathed heaters have been used as safe electrical sources of heat in many applications—most prominently in home appliances, such as toasters, irons, hot water heaters, ovens, and ranges. We have come to take Calrods (General Electric trade mark) for granted. Even though they are simple in construction, many complex physical and chemical characteristics have to be accommodated both during fabrication and during service life.

The Calrod is a spiral resistance heater mounted in a stainless-steel sheath. A granular ceramic oxide, conventionally MgO, supports the heater wire mechanically and both isolates it electrically from the sheath and connects it thermally to the outside world. Special equipment aligns the spiral element centrally in the sheath and at the same time fills the space inside the sheath with the MgO powder. The filled unit is then mechanically swaged to a smaller diameter to compact the powder and shape it as a spiral cooktop elements or other shape. The surface on which the cooking pot rests is generally flattened to improve heat transfer. During all these swaging operations, the separation between the heater wire and the sheath is maintained by the granular powder. MgO is a perfect material for this application. It is soft enough not to damage the metal components during swaging; it is a cubic material and can, therefore, readily shear to allow compaction. MgO has a high thermal conductivity; its high thermal expansion coefficient minimizes separation from the sheath during heater operation. Moreover, MgO is a good electrical insulator.

Chemists, ceramists, metallurgists, and engineers have worked to combine in this device the rapid heat-up and cool-down features of gas heat with the cleanliness and simplicity of electric heat. Contrary to popular belief, the Calrod tubular sheathed heater will boil a quart of water on a range top faster than a gas-flame unit.

Research on sheathed heaters begun in 1910 by Chester Moore of the GE Research Laboratory resulted in a 1913 U.S. Patent (1,077,734) related to straight heater wire units. A spiral wound heater was patented in the United States in 1921 (1,367,341) by C. C. Abbott.

Some of the physical properties of the MgO fill had to be tailored for successful operation of tubular-sheathed heaters. The minimum diameter of early units was limited by electrical leakage from the heater wire to the sheath. L. Navias and O. G. Vogel of the GE Research Laboratory addressed the leakage issue and determined that specific amounts of SiO_2 added to the MgO powder would tie up free CaO and reduce the electrical leakage (U.S. Pat. No. 2,285,953, June, 1942). SiO_2 additions are now an industry standard. Thermal conductivity of the MgO powder also was an issue. Fine powder acts as a poor thermal conductor. To improve the heat transfer qualities of the electrical insulation, the grain size of the MgO is increased by converting magnesite ($MgCO_3$) in a carbon-arc furnace to molten MgO (>2800°C) and allowing the MgO to cool to large, single crystals, which are crushed into coarse granular MgO.

A variety of chemical processes occur in an energized Calrod. They include the consumption of both oxygen and nitrogen by the hot-metal components. Oxygen and nitrogen are resupplied by permeation through a porous or semipermeable end cap and through the interconnected porosity of the granular medium. The presence of the gas in the interior of the unit actually contributes to the heat transfer. If an air-filled unit is hermetically sealed, the heater wire runs too hot and melts in a short period of time because of the loss in thermal conductivity between the heater and the sheath resulting from consumption of the internal gas by the hot-metal components.

The gas–metal reactions within the unit can result in extremely low partial pressures of both oxygen and nitrogen. The electrical properties of commercial ceramic oxides, particularly MgO, can be quite sensitive to oxygen partial pressure. The recognition and understanding of the complex relationship between the chemical reactions in an energized Calrod and its physical properties have led to improved unit design and better manufacturing controls.

Single-Crystal Oxide Materials

by Tony Keig

Single-crystal synthetic ruby had become an important bearing material at the beginning of World War II, and, in order to protect its strategic interests, the U.S. government commissioned the Linde Division of Union Carbide to develop a domestic source using the high-temperature Verneuil process that was being used commercially in Europe. It was to this established source that scientists at Bell Laboratories turned during the early 1960s to test the newly developed theories for solid-state ruby lasers.

The small size and poor optical quality of the available single crystals limited the scope of these early experiments, and material scientists were given the challenge of developing alternative methods for reliably producing the required large-sized, high-quality products. The best success came with the Czochralski (CZ) technique, where a high-temperature melt was established using ultrapure raw materials, and single crystals were pulled from the melt under carefully controlled conditions. The technique had been known for many years and used with some success under laboratory conditions to produce small-sized, lower-melting-temperature metals and compounds. Extending the technique to the much-higher oxide-melting temperatures and building furnace structures capable of commercially producing single crystals up to 3–4 in. in diameter presented a much more difficult challenge, especially where the science for achieving this was and continues to remain largely empirical. Union Carbide Corporation, under the technical leadership of Fedia Charvat and supported by the cooperative scientific resources at Bell Laboratories, remained a leader in developing this commercial melt–growth single-crystal process.

Ruby was the first successful solid-state laser, and it found its main use through a military need for tank range finders. Neodymium-doped yttrium aluminum garnet (Nd:YAG), which provided both a continuous and pulsed source of laser radiation has proved to be a more successful commercial product and is currently used in many machining and medical applications. Equipment is available for drilling, welding, cutting, and shaping metals and ceramics. In the medical area, procedures using YAG lasers have been reported for cataract, skin, and internal cancer surgery. Other electrooptical single crystals, notably lithium niobate, are starting to find increasing application in switching and modulating fiber-optic signals.

Applications for oxide single crystals have not been restricted to lasers. The less-expensive Verneuil method was developed to a production method for synthetic rubies, sapphires, and star sapphire. Perfect emeralds have been made using the hydrothermal method. Not so perfect, but commercially more successful emeralds have emerged using the flux process. Many diamond substitutes have been attempted, the most realistic being cubic zirconia, which uses a modified CZ process. Applications have been studied for undoped sapphire ranging from transparent armor to ultraviolet windows. The most successful development has been a substrate for an epitaxial silicon film from which radiation-hardened semiconductor devices are fabricated.

Mass Production of Refractory Oxide Crystals: Cubic-Zirconia

by Joseph F. Wenckus, Zirmat Corporation

Single-crystal cubic-zirconia (CZ) is the most widely accepted diamond simulant available in the gem and jewelry industry today. Since the introduction of CZ gems in 1977, crystal production has grown dramatically, with present annual output estimated to be well in excess of 400 metric tons (>2 billion carats).

Crystal growth by solidification from a contained melt of the desired composition is by far the most rapid and economical method used to produce large crystals. Unfortunately, there is no known crucible material that can be used to contain molten zirconia (melting point ~2750°C) under oxidizing conditions. Since the early 1960s a variety of methods have been explored to produce single-crystal CZ: solution methods using molten fluxes, plasma, and arc melting—and even a large solar furnace in the French Pyrenees—have been used to prepare tiny crystals of CZ for research applications. In 1969 the French researchers Roulin, Vitter, and Deportes were the first to report the growth of relatively large (15 mm long) yttria-stabilized cubic-zirconia crystals using a new technique called skull-melting.

The skull-melting process may be thought of as a very-high-temperature analog of microwave cooking, where the microwave energy heats the contents in the dish, but not the dish itself. Skull-melting utilizes a copper crucible-like structure (consisting of individual water-cooled tubes adjacent to one another) to surround the radio frequency-heated molten zirconia, which is contained by a thin, skinlike shell or "skull" of its own composition (see Fig. 38). Thus the contained melt is not in contact with, or contaminated by, the cold crucible. The melt then is cooled slowly from the bottom up, thus forcing columnar CZ crystals to nucleate at the base of the melt. As the crucible is slowly lowered below the radio-frequency coil, directional solidification of the crystals proceeds upward to the melt surface until the molten charge is converted to a mass of elongated CZ crystals. The growth process, which is illustrated in the attached figure, can be completed in several days.

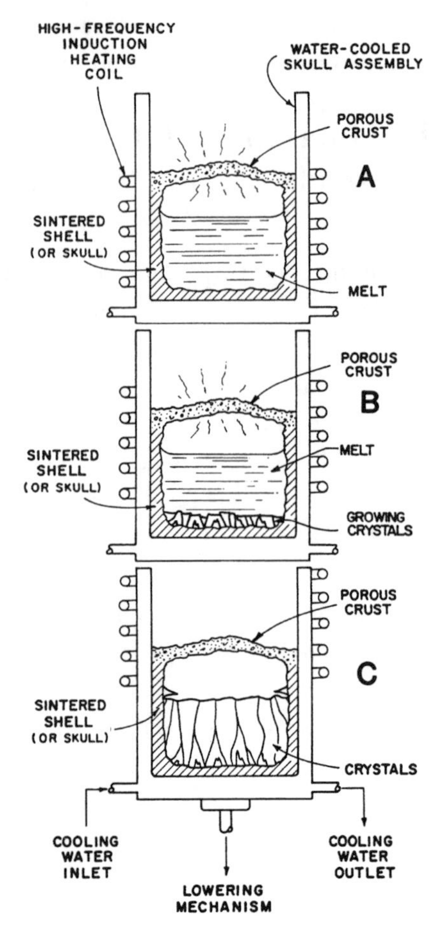

Figure 38. *Skull-melting crystal growth process showing lowering of assembly and accompanying crystal growth. Courtesy of Zirmat Corporation.*

In the early 1970s scientists at the Lebedev Physics Institute (Moscow) expanded upon this new technique in their on-going research for new laser crystals. Although doped CZ crystal did in fact exhibit laser characteristic, their performance was disappointing and further research on the laser applications of CZ was abandoned. However, the researchers at Lebedev were the first to recognize the unique diamond-like optical properties of cubic-zirconia and a new diamond simulant was born.

During the past 25 years, skull-melting technology has undergone dramatic development in the transition from laboratory-scale crystal growth in an early skull-melting furnace operating with a 30 kg melt. Today, skull-melting furnaces containing zirconia melts well in excess of 2 metric tons are now operating of a routine basis.

Market forces have, in fact, driven this rapid technological evolution. In 1977, the average selling price for as-grown CZ crystals to the gem trade was $2500/kg or $0.50/ct. Despite the fact that crystal size and quality have improved dramatically since that time, the current selling price is less than $50/kg ($0.01/ct.). Indeed, CZ crystals have become a low-cost commodity material for the gem trade, and efficient, large-scale crystal production is absolutely essential for profitable operations.

Gas–Metal Eutectic Direct Bonding for Advanced Metallization and Metal Joining

by Victor A. Greenhut, Rutgers University

The gas–metal eutectic method is one of the most rapidly growing technologies for metallizing ceramics and joining them to bulk metals. It is a direct-bonding technology in which no intermediary bonding or adhesion-promoting materials need be introduced between ceramic and metal. Currently, the most common application is direct bonding of copper metallization with alumina, often termed "direct-bonded copper" or "DBC" bonding, for a variety of sophisticated hybrid microelectronic circuits for applications such as cell phones, computers, automotive microprocessors, power hybrids, and numerous electronic-packaging applications. This method is growing very rapidly because it provides a high-quality, strong, reliable, economical bond without intermediary material, which interferes with performance. It is particularly useful in high-power circuitry that requires the thermal and electrical advantages of pure copper directly on alumina without intermediary materials. Defect-free bonds can be produced reliably with bond strengths exceeding that of the ceramic.

The key feature of this technology was discovered and patented by Burgess and Neugebauer at General Electric Company almost three decades ago, but it was first commercialized significantly only in very recent years. The phase diagram for copper and oxygen (see Fig. 39) reveals that copper with a small amount of oxygen melts at a lower temperature than copper metal itself. Thus, a thin layer of copper–oxygen eutectic melt can form without the bulk of copper metal melting. A very useful application of the process involves additive electroplating to pro-

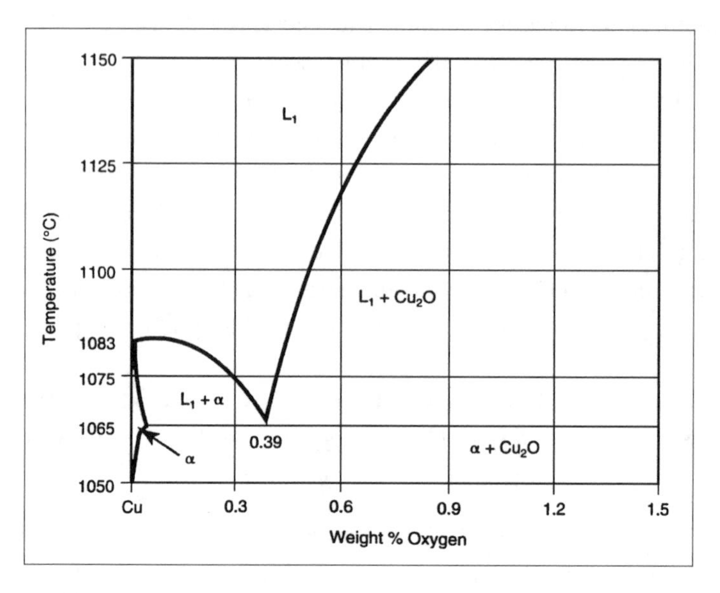

Figure 39. *Copper–oxygen binary phase diagram showing depression of the melting point of copper by a small amount of oxygen addition. Courtesy of Victor A. Greenhut, Rutgers University.*

duce complex copper circuit patterns on ceramic for hybrid microelectronics. An example is shown in figure 2. Patterns are defined photolithographically that readily result in sharp line definition of repeatable patterns capable of high-volume production.

It is expected that the technology will expand significantly into many other materials and applications during the next five years. The Ceramic–Metal Systems Group at Rutgers University has researched and developed major process improvements, a fundamental understanding of bond mechanisms, the effect of bonding technique and/or composition, and extension to new materials and applications. The understanding gained in basic studies on bonding copper to alumina has provided Cirqon Technologies, Inc., with process refinements that improve bond strength and that have lowered rejection rates by more than 80%. Technology transfer of methods for bonding of copper and aluminum nitride has been conducted internationally with Dow Corporation, a leading supplier of aluminum nitride powders. Current research activity concentrates on high-temperature structural ceramics (silicon nitride and silicon carbide) and metals (nickel alloys), in cooperation with United Technologies, Kyocera Industrial Ceramics, Cirqon Technologies, Inc., and others.

Ceramics for Paper

by Victor A. Greenhut, Rutgers University

Most paper has 2%–10% ceramic as a so called filler material; some papers have even more. The ceramic is actually a very important active ingredient providing paper with opacity and whiteness, while controlling the flow of ink in writing and printing. Without the ceramic, ink would be absorbed by and smear into the paper. The ceramic also can provide color. Ceramic has been used for many centuries as an important paper additive, with china clays, such as kaolin, having very long histories as the applied material.

During the past century, titania (titanium dioxide) has displaced much kaolin usage. Titania has a much higher light scattering than kaolin, coupled with very fine particle size, which makes it ideal for whitening, brightening, and opacifying paper. Titania also has been used extensively because of these properties as the white base for oil-type paints and as cosmetic whiteners in products such as toothpaste. In an effort to regain the very sizable paper market (and add value), kaolin producers began to calcine the kaolin clay. The clay is heated to high temperature in a rotary kiln. The individual clay platelet particles release structural water and collapse to form soft metakaolin glass platelets. With continued heating, metakaolin eventually transforms to mullite crystals with similar whitening and opacifying properties to titania. Unfortunately, the calcined clay was more abrasive for the paper forming and handling machinery. Studies at Rutgers University solved the problem producing a low-abrasion calcined kaolin product that also was whiter, brighter, and more opacifying.

It was found that a poorly understood heat release (exotherm) at ~980°C (1800°F) occurred when many submicroscopic, nanoscale crystals began to form within each of the microscopic metakaolin platelets. This could only be seen with a transmission electron microscope at very high magnification. The nature of this high-temperature transformation depended on the rate of heating, gas environment, and trace additives. Use of small amounts of mineralizer additives and water vapor was found to provide engineering control of the formation of these tiny particles encapsulated in the metakaolin glass. The small particles could give paper the desired whiteness, brightness, and opacification, while encapsulation in a glass made a relatively nonabrasive powder. The technology resulting from the study was commercialized. After about a decade, the resulting calcined kaolin powders have virtually displaced titania as the paper additive of choice for whitening, brightening, and opacification with low abrasion for paper-making machinery.

Coatings

Flame-Sprayed Ceramic Coatings

by Neil Ault, Saint-Gobain/Norton

In 1952 H. F. G. Ueltz and Max Wheildon at Norton Company Research Laboratories were investigating methods of making surfaces more wear resistant, and they wanted to use the hardness of ceramics. Attempts had been made to produce ceramic coatings of refractory materials by flame spraying prior to 1950, but these had resulted in soft, porous coatings with poor adherence. Ueltz and Wheildon recognized that the flame spraying of ceramic powders with an oxypropane or oxyacetylene torch resulted in such short retention time in the flame that little melting of the powders occurred. To ensure that the ceramic was melted, they put a ceramic rod into an oxyacetylene flame until melting occurred, and used a blast of air around the flame to atomize the molten droplets. The blast of air projected the spray against a substrate, causing the droplets to flatten, thus producing hard, adherent coatings. Then came the effort to design a flame spray gun through which rods could be fed, melted, and sprayed. Ceramic rods were developed that could withstand the thermal shock involved without spalling. After the initial success with refractories, such as alumina and zirconia, other materials also were made as coatings, the most important of which was chromium oxide, which could be made with an open porosity of $<3\%$.

The keys to the quality of the coatings were the total melting of the ceramic in the flame, the velocity imparted to the molten particle so that it flattened against the surface being coated, and rods that could feed without spalling.

These unique coatings (trade name ROKIDE) were used in some of the earliest U.S. rocket programs to provide the erosion resistance, thermal insulation, and refractoriness required for rocket propulsion nozzles. They also have found use as electrical insulation in small motors. The major applications today take advantage of the wear resistance of the chromium oxide coatings (original goal of the research) in applications in the petrochemical industry for pump parts; in the paper industry for sleeves, shafts, suction box covers, etc.; in the wire-drawing industry for capstans, sheaves, and guides; and in the textile and packaging industries for various wear-resistant parts.

Figure 40. *Flame spray gun for applying ceramic coatings. Courtesy of Norton Pakco Industrial Ceramics.*

Authors Brief Biographies

Ainsworth, John H.

John H. Ainsworth obtained his B.S., M.S., and Ph.D., all in ceramic engineering, from the University of Missouri-Rolla. He has worked for Babcock and Wilcox, Bethlehem Steel, Chromalloy, and General Refractories. He is currently employed with North American Refractories Company. His interests include mechanical and thermal properties of ceramic materials.

Alspaugh, Robert

Robert Alspaugh received his B.S. in chemistry from the State University of New York at Cortland, NY. He spent more than 20 years in manufacturing glass for lighting products in a wide variety of positions, and is now Plant Manager of the Wellsboro, PA, Osram Sylvania, Inc., glass plant.

Araujo, Roger J.

Roger J. Araujo received his Ph.D. in physical chemistry from Brown University in 1962 and went directly to work for Corning Incorporated. He was made Manager of Photochemical Research in 1970 and made Fellow in 1986. He has 32 patents and more than 50 publications, many of them related to photochromic and photosensitive glasses.

Ault, Neil N.

Neil N. Ault received his B.S., M.S., and Ph.D. in ceramic engineering from The Ohio State University. He was employed by the Norton Company in research, retiring as Technical Director, Refractories Division, in 1987. Since then, he has been an independent consultant in the field of high-temperaure advanced ceramics.

Baader, Felix H.

Felix H. Baader received his materials science diploma in 1991 and his Ph.D. from the ETH Aurich in 1995. He now is Senior Research Scientist at Tegimenta AG, Switzerland.

Beall, George H.

George H. Beall holds degrees (B.S., M.S.) in physics and geology from McGill University in Montreal. He joined Corning after receiving his Ph.D. in geology from the Massachusetts Institute of Technology in 1962. He has conducted investigations in the areas of transparent glass-ceramics, machinable glass-ceramics, strong and tough glass-ceramics, low-temperature glasses for glass–polymer blends, low and negative coefficient of thermal expansion materials, and photosensitive and polychromatic glasses. His continuing interest involves the relation between microstructure and properties of multiphase glass-derived materials. He is a Research Fellow in the Science and Technology Division of Corning Incorporated. He has published more than 70 technical papers, edited one book, and holds 81 patents. He is a Courtesy Professor in the Department of Materials Science and Engineering at Cornell University. Beall is a Fellow of The American Ceramic Society and has received the Morey Award in 1988, the John Jeppson Award in 1993, the Samuel Geijsbeek Award in 1993, and delivered the Edward Orton Memorial Lecture in 1996. He also has been awarded the American Chemical Society's Chemistry of Materials Award in 1995 and the Corning Section's Sullivan Award in 1997.

Beck, Warren R.

Warren R. Beck received his B.S. in ceramics from The Pennsylvania State University. He pursued graduate study in ceramics, also at Penn State, and continued graduate studies in geology and physical chemistry at the University of Minnesota. He was on the ceramics research staff at Penn State for two years, then was recruited by Nelson W. Taylor when Taylor resigned as head of the Ceramics Department at Penn State to direct ceramics research at the 3M Company in Saint Paul, MN. Beck spent the balance of his professional career in research at 3M. While there his chief involvement was in the development of high-refractive-index glass microspheres and hollow glass microspheres. He was a Corporate Scientist there when he retired and is now doing occasional consulting in the field of his technical interests, which has broadened to include the general area of particle morphology.

Boccaccini, Aldo R.

Aldo R. Boccaccini is currently Research Associate in the Materials Research Group, Technical University of Ilmenau, Germany. He took this position recently after having been a visiting researcher at the Institute for Mechanics and Materials, University of California, San Diego, and at the School of Metallurgy and Materials, University of Birmingham (U.K.). Originally from Argentina, he earned his Ph.D. in materials science (glass and composite materials) from the Technical University of Aachen (RWTH), Germany, in 1994. His research interests include the processing and mechanical properties characterization of glass and ceramic-matrix composites, and the development of new glass-ceramics by recycling industrial waste.

Boch, Philippe

Philippe Boch was born in 1938. After studies of mechanical engineering in Poitiers (France) and a doctorate thesis in physical metallurgy, he moved to the University of Limoges. For eight years, he was director of the School of Ceramics, the only French college of engineering uniquely devoted to ceramics. In 1988, he moved to Paris, where he is professor of materials science and solid-state chemistry at both Pierre & Marie Curie University and the School of Industrial Physics and Chemistry. Philippe Boch has been the president of the materials science program in CNRS (which may be compared to NSF) for eight years, and he is the past president of the French Society for Metals and Materials. His current scientific interests are cementitious materials, in particular those for trapping chemical or nuclear pollutants. He is author or coauthor of more than 160 scientific papers.

Borom, Marcus P.

Marcus P. Borom received a B.S. in ceramic engineering from the Georgia Institute of Technology in 1956 followed by a year of postgraduate study at the Swiss Federal Institute of Technology in Zurich. He served as a ceramic engineer for the Ferro Corporation from 1956 to 1961, working in the area of porcelain enamels. In 1961 he entered graduate school at the University of California at Berkeley, where he received a Ph.D. in materials science in 1965. He joined the General Electric Research and Development Center in 1965, where he worked until retirement in 1998. At the Center he conducted research in the following areas: wetting, reaction, and bonding in glass–metal and ceramic–metal systems; kinetics of high-temperature, solid–liquid–gas reactions; glass passivation of semiconductor devices; advanced heat sink materials; thermal protection systems for the space shuttle; mold and core materials and processes for casting of DS eutectic superalloys; innovative sintering processes for the fabrication of cutting tool materials; advanced, fiber-reinforced, ceramic-matrix composite materials; nozzles for spray atomization of superalloys; and processing, microstructure, and properties of thermal barrier coatings. Borom also served 18 months as Technical Coordinator at the laboratory level and three years as manager of the Cutting and Wear Resistant Materials Program. Borom is a member and Fellow of The American Ceramic Society. He has authored 40 technical papers and holds 37 patents.

Boulos, Edward N.

Edward N. Boulos received a B.S. from Cairo University in chemistry and physics, an M.S. from the American University in Cairo, and a Ph.D. from the University of Missouri-Rolla. He also received a professional degree in ceramic engineering from the University of Missouri-Rolla. He was Senior Research Scientist, Anchor Hocking, from 1980 to 1984. He was then Staff Technical Specialist, Ford Motor Company, from 1984 to 1997, and is now Staff Technical Specialist, Visteon Glass Systems (an enterprise of Ford Motor Company). His technical interest is the chemistry and physics of glass and glass coatings.

Bradt, Richard

Richard Bradt graduated from the Massachusetts Institute of Technology with a B.S. in metallurgy, worked in industry with Fansteel Metallurgical Corporation in North Chicago, and then returned to graduate school at Rensselaer Polytechnic Institute for his M.S. and Ph.D. in materials engineering. He has served on the faculties at The Pennsylvania State University, the University of Washington, the University of Nevada, and the University of Alabama. He has published more than 300 manuscripts, advised nearly 100 graduate theses, and is a Distinguished Life Member of UNITECR, the international refractories organization.

Burggraaf, A. J.

A. J. Burggraaf was a Professor of Inorganic Materials Science at the University of Twente, Enschede, The Netherlands, from 1970 until his retirement in 1994. Preceding his last position he joined the Philips Electronic Company from 1960 to 1970 and the Technical University of Eindhoven from 1958 to 1960. His active research fields were glass and glass-ceramics, ferroelectric materials, electrical- and oxygen-ion-conducting ceramics, tough and strong zirconia/alumina ceramics, ceramic membranes, and nanostructured ceramics. He published about 400 papers, holds 20 patents, and edited several books. He served the government and professional organizations in many functions. He holds three professional awards and was distinguished by the government in 1994.

Burke, Joseph E.

Joseph E. Burke received his B.A. in chemistry from McMaster University and his Ph.D. in chemistry from Cornell University. After other jobs, he was a group leader at Los Alamos National Laboratory through World War II, and was an Assistant then Associate Professor of Metallurgy at the Institute for the Study of Metals at the University of Chicago. From 1949 to 1954 he was with the General Electric Knolls Atomic Power Laboratory, finally as Manager of Metallurgy. In 1954 he organized and then managed for many years the Ceramic Studies Section at the GE Corporate Research and Development Center. In 1979 he retired. In recent years his technical interests have been with the factors controlling the development of microstructure in polycrystalline materials, and in particular with the mechanisms and kinetics of the important reactions.

Bursey, Roger W., Jr.

Roger W. Bursey Jr. received a B.S. in mechanical engineering from Florida Atlantic University. He is a Project Engineer at Pratt & Whitney, West Palm Beach, FL, where he is responsible for the development of components for the space shuttle main engine high-pressure turbopumps. While at Pratt & Whitney, he has been responsible for developing high-speed bearings that utilize Si_3N_4 balls and rollers.

Cameron, Neil

Neil Cameron received his B.S. in mechanical engineering from the University of Wales and his M.S. and Ph.D. in theoretical and applied mechanics from the University of Illinois. He is a laboratory manager at Owens Corning, where he has worked since 1980. Prior to this he worked at Pilkington, PPG Industries, and, as an apprentice, Rolls-Royce. His research interests cover most aspects of glass-fiber manufacture and performance.

Carruthers, David

David Carruthers received a B.S. in mechanical engineering from Virginia Polytechnic Institute. He spent 25 years in the gas turbine industry at both Pratt & Whitney Aircraft and AlliedSignal Engines, where he supported and managed programs aimed at developing ceramic, ceramic composite, and carbon–carbon components for gas turbines. In 1991 he joined Kyocera Industrial Ceramics Corporation (KICC), Vancouver, WA, where he is the Strategic Marketing Manager and is responsible for developing new business opportunities for structural ceramics through development contracts. He also manages KICC's test and inspection laboratory.

Cascone, Paul J.

Paul J. Cascone is Senior Vice-President of Technology for The Argen Corporation, one of the top three manufacturers of dental alloys worldwide. He holds a B.E. in metallurgical engineering from New York University, an M.S. in ceramic science from Rutgers University, and has attended the Harvard Business School AMP program. Previously Mr. Cascone was President of J. F. Jelenko and Company, a supplier of dental materials.

Cava, Robert J.

Robert J. Cava graduated with a Ph.D. in ceramics from Massachusetts Institute of Technology in 1978. After a one-year postdoctoral fellowship at the National Bureau of Standards with R. S. Roth, he joined the staff of Bell Laboratories. He was named a distinguished member of technical staff there in 1985. In 1996 he joined the faculty at Princeton University in the Chemistry Department and the Princeton Materials Institute. He was recently awarded the Mattias prize for the discovery of new superconducting materials, and the Chemistry of Materials prize by the American Chemical Society. His interests are in the synthesis, crystallography, and phase equilibria of new transition-metal oxides, pnictides, and intermetallic compounds with unusual electronic and magnetic properties.

Danielson, Paul

Paul Danielson received his undergraduate degree in chemistry from Franklin and Marshall College, and a Ph.D. in inorganic chemistry from the University of Connecticut in 1974. After a postdoctoral in chalcogenide glasses at the Vitreous State Laboratories at the Catholic University of America, he joined Corning Incorporated (then Corning Glass Works) in 1976, where he has been doing glass composition and processing R&D ever since. He holds 17 U.S. patents.

Davis, Allen D., Jr.

Allen D. Davis Jr. obtained a B.S. in ceramic science at The Pennsylvania State University in 1970 and a Ph.D. in ceramic science from The Pennsylvania State University in 1977. His professional career began at Corning Glass Works in Corning, NY, in 1974 where he was a Senior Ceramist working with catalytic converter substrates and zirconia ceramics. In 1977 he moved to Corhart Refractories in Louisville, KY, where he has held various professional and management positions in product development, laboratory and technical services, and sales and marketing. He is currently Manager of Applications Engineering. He holds one patent, has published many technical articles, and has written chapters on glassmelting furnaces and refractories for *The Encyclopedia of Materials Science and Engineering* and for the *The Handbook of Glass Manufacture*. He is a member of Keramos, the Refractory Ceramics Division of The American Ceramic Society, and of ASTM Committee C8 for refractories.

Deeg, Emil W.

Emil W. Deeg was born, raised, and educated in Germany. He holds a Physics Diploma and a Dr. rer. nat. from the Julius-Maximilians-

Universitaet with thesis work at the Max-Planck-Institut fuer Silikatforschung, Wuerzburg. He did postdoctoral work at Bell Telephone Laboratories, Allentown, PA, and served subsequently as Director of Scientific Development and Metrology at the Jenaer Glaswerke Schott & Gen. in Mainz, Germany. In 1965, Dr. Deeg was appointed Associate Professor of Physics and Solid State Science at the American University in Cairo, Egypt, and also lectured on molecular engineering at Alexandria University. He moved to the United States in 1967, where he served international glass and electronic companies in various management positions. In 1992 he retired from AMP Inc., Harrisburg, PA, but continues to work for the company as consultant and editor of the *AMP Journal of Technology*.

Dejneka, Matthew J.

Matthew J. Dejneka earned his Ph.D. from Rutgers University in 1995, where he worked on the sol–gel synthesis of fluoride glasses and fiber, as well as the characterization of rare-earth-doped materials. He held internships at Argonne National Laboratory and Garrett Ceramic Components, where he worked on high-temperature energy cells and pressure casting of silicon nitride turbines. Upon completion of his degree, he joined Corning Incorporated, where he invented transparent LaF_3 glass-ceramics. He is a Senior Research Scientist in the Glass and Glass-Ceramics Core Technology Group at Corning. He currently works on novel fiber fabrication methods, and glass composition for such applications as optical amplifiers, three-dimensional upconversion displays, LCD displays, lighting, and photonic devices.

DeVries, Robert C.

Robert C. DeVries received his B.A. from DePauw University and his Ph.D. from The Pennsylvania State University, both in geochemistry and mineralogy. He was Staff Scientist at the General Electric Research Laboratory from 1954 to 1961, Associate Professor of Materials Science at Rensselaer Polytechnic Institute from 1961 to 1965, and Staff Scientist at the General Electric Research and Development Center from 1965 to 1988. He is a consultant and Adjunct Professor at The Pennsylvania State University. His interests are in the synthesis and properties of ceramic materials; relationships of both crystal structure and microstructure to properties; crystal chemistry; phase equilibria of inorganic, nonmetallic systems at both high- and low-pressure conditions; crystal growth; synthesis of diamond and cubic BN at high and low pressure; origin of natural diamond; characterization of synthesized and natural diamond; and gemology. He is a member of the National Academy of Engineering.

diBenedetto, Bernard A.

Bernard A. diBenedetto received his B.S. in ceramic engineering from Alfred University in 1958. Since then he has worked on the development of materials at Raytheon Company and Tyco Laboratories, Inc. His work included process development of piezoelectric compositions, microwave dielectric ceramics, IR and radiation detector materials, and IR window and lens materials. His early work centered on crystal growth and more recently he has been Production Manager for Raytheon's family of new chemical vapor deposition IR window materials.

Eicher, Robert J.

Robert J. Eicher graduated from Carnegie-Mellon University with a B.S. in mechanical engineering. He is employed by Swindell Dressler International Company, Pittsburgh, PA, as Senior Project Manager. He is a Past President of the Pennsylvania Ceramic Association.

Fisher, Robert E.

Robert E. Fisher received his B.S. in chemistry from Grinnell College in 1961 and his M.S. in metallurgical engineering from the Illinois Institute of Technology in 1964. He worked from 1973 to 1994 at Plibrico Company in various positions in R&D, then as Technical Director, then as Managing Director of Plibrico USA, and then President of Plibrico. Since April 1994 he has been a consultant in monolithic refractories technology. He has been a vice president of The American Ceramic Society, vice president of the inaugural UNITECR Congress in 1989, past Chair of the Refractory Ceramics Division, Fellow of both ACerS and ASTM, editor of two ACerS books on monolithic refractories technology, author/coauthor of more than 25 technical papers, and holds seven patents. His principal technical interests are the development and testing of advanced monolithic refractories and encouraging their use to replace brick.

Freiman, Stephen W.

Stephen W. Freiman received a B.Ch.E. and an M.S. in metallurgy from the Georgia Institute of Technology. He received a Ph.D. in materials science and engineering from the University of Florida. Freiman worked at the IIT Research Institute until 1971, when he joined the Naval Research Laboratory. He began work at NIST (then NBS) in 1978 where he is now Chief of the Ceramics Division. Freiman has published more than 150 scientific papers focusing on the mechanical properties of brittle materials, and has contributed to many books and monographs. Freiman is a Fellow of The American Ceramic Society (ACerS). He is a former vice president, Treasurer, President-Elect, and the current (1998) President of ACerS. He is the U.S. representative to the steering committee of VAMAS (the Versailles Project on Advanced Materials and Standards).

Gauckler, Ludwig J.

Ludwig J. Gauckler is Professor at ETH Zurich, Switzerland. His research interests include the use of colloidal chemistry for ceramics and thin films as well as the application of computer-assisted thermodynamic modeling in processing of high-temperature superconductors, fast-ion conductors for solid-oxide fuel cell systems, and structural ceramics. He holds a diploma in solid-state physics from the University of Stuttgart and a D.Sc. from the Max Planck Institute for Materials Science at Stuttgart.

Geczi, Linda S.

Linda S. Geczi received her B.S. and M.S. in ceramic engineering from Rutgers University. Her master's thesis dealt with environmental laws affecting the use of lead in the ceramic industry. She is a member of the staff of the McLaren Center for Ceramic Research at Rutgers University, where she has participated in research on lead glazes.

Goldman, Alex

Alex Goldman received his M.A. and Ph.D. from Columbia University. He is President of Ferrite Technology Worldwide, a Pittsburgh-based ferrite consulting company. For the prior 21 years, he was Corporate Director of Research for Spang and Company and its Magnetics Division (a large ferrite and other magnetic component manufacturer). Previous to that, he was Senior Research Scientist in the Magnetics Department of the Westinghouse Research Laboratories. His major interests are high-permeability and high-frequency power ferrites, microstructural–magnetic correlations, nonconventional ferrite processing, and nanocrystalline materials. He is the author of *Modern Ferrite Technology* and his book *Handbook of Modern Ferromagnetic Materials* was published in 1998 by Kluwer Academic Publishers. He is a Fellow (emeritus) of The American Ceramic Society and a life member of the IEEE and the American Chemical Society. He was awarded the Takeshi Takei Award for Ferrite Achievements at ICF7 in Bordeaux in 1996. He was cochairman of ICF4 (International Conference on Ferrites) in 1984 in San Francisco.

Gordon, Ronald

Ronald Gordon received a B.S. in chemical engineering at the University of California (Berkeley) in 1959. He received an M.S. in ceramic engineering at the University of California (Berkeley) in 1961 and a Sc.D. in ceramics at the Massachusetts Institute of Technology in 1964. He was a Professor of Materials Science and Engineering at the University of Utah between 1964 and 1989. He has served as Head, Department of Materials Science and Engineering, Virginia Polytechnic Institute and State University, since 1989. In 1977 he cofounded Ceramatec, Inc., in Salt Lake City, UT, and served as its Chairman and CEO until 1988. He was employed in the General Electric Lighting Research Laboratory at Nela Park between 1971 and 1973. His research interests include ceramic materials processing and characterization, microstructural development, diffusional creep processes in polycrystalline ceramics, ionic conduction and diffusion, and electrochemical applications of ceramic materials.

Graule, Thomas J.

Thomas J. Graule is Senior Scientist at Katadyn Product AG, where he is responsible for product and process development in ceramics for filtration and catalyst carriers. His chemistry degrees include an M.S. from the Engineering College of Aachen, Germany, a Diploma from the University of Siegen, Germany, and a Ph.D. from the Max

Planck Institute for Metals Research in Stuttgart. He joined L. J. Gauckler in 1988 as a postdoctoral research assistant at the Swill Federal Institute of Technology in Zurich, where he led the colloidal chemistry group as the Chair of Nonmetallic Materials.

Greenhut, Victor A.

Victor A. Greenhut earned his B.S. in physics from City College of New York (1965) and Ph.D. (1970) in mechanics and materials science (minor in ceramics) from Rutgers University. He conducted postdoctoral research at Oxford University (1970–1971). He is the Corning/Saint-Gobain—Malcolm G. McLaren Distinguished Chair in Ceramic Engineering. He is Executive Officer of the Department of Ceramic and Materials Engineering, Rutgers—The State University of New Jersey. He served previously as Associate Dean and Director of Interdisciplinary Studies for the Graduate School (1989–1993). Professor Greenhut's research concentrates on ceramic–metal systems, ceramic-matrix composites, ceramic/glass-to-metal joining, bioengineering materials, traditional ceramics, and structure–property relationships. He has extensive practical research, consulting, and workshop experience with industry, military, and government in the areas of manufacture and properties of ceramic–metal systems, composites, ceramic/glass-to-metal joining, manufacturing, materials characterization and properties, and failure analysis.

Greskovich, Charles D.

Charles D. Greskovich received his B.S. in ceramic technology in 1964 and his M.S. and Ph.D. in ceramic science in 1966 and 1968, respectively, all from The Pennsylvania State University. He was awarded an NSF postdoctoral fellowship in 1968 and studied at Clausthal Technical University in Germany. In 1969 he became a staff ceramist at GE's Corporate R&D Center in Schenectady, NY. Greskovich's research has been focused primarily on the influence of powder preparation, processing, and microstructure on the properties of advanced high-technology polycrystalline ceramics, and his work has been widely recognized in the form of numerous national, international, and company awards. He has published more than 50 papers and been awarded 46 U.S. patents.

Griffin, Don

Don Griffin received his B.S. in ceramic engineering from Virginia Polytechnic Institute, Blacksburg, VA, in 1976 and M.S. in materials engineering in 1978. From 1977 to 1980 he was employed as a Research Engineer by Taylor Refractories in Cincinnati, OH. Since 1980 he has been employed by Baker Refractories in various capacities. These have included Technical Service Engineer, Research Engineer, Director of Research and Development, and, most currently, General Manager—Baker Engineered Ceramics. His main technical interests have been focused on the development and application of doloma refractories. He is a member of ASTM, AISE, and The American Ceramic Society.

Gross, Karlis A.

Karlis A. Gross graduated from Monash University, Australia, with a B.E. (1987) and M.Eng.Sci. (1990), both from the Department of Materials Engineering, where his focus was on bioceramic coatings. In 1991, he spent a year at the Riga Technical University, Latvia, continuing to work in his field but concentrating more on hydroxyapatite powder synthesis and glass technology. As a recipient of a DAAD scholarship, he spent five months at the Technical University of Aachen, Germany, on preparing this powder in a suitable form for coating by thermal spraying. Further studies commenced at The State University of New York at Stony Brook, NY, in 1992 to solve an intrinsic problem of plasma-sprayed hydroxyapatite coatings. Three years later, he received a Ph.D. in the Department of Materials Science and Engineering, where he determined the cause of amorphous phase formation with hydroxyapatite during the thermal-spray operation. Postdoctoral appointments have taken Gross to the Technical University of Tampere, Finland, and the University of Limoges, France, followed by a joint position at the University of Technology, Sydney, and Monash University, where his work on the development of hydroxyapatite coatings is funded by the Australian Research Council. His main interests relate to biomaterials, coating technology, and pollution control from engineering processes.

Gururaja, Raj

Raj Gururaja received his B.Sc. and M.Sc. in physics from the University of Mysore, India (in 1974 and 1976, respectively), and a

Master of Tech. in materials science from the Indian Institute of Technology, Kanpur, in 1978. He received his Ph.D. in solid-state science from The Pennsylvania State University in 1984, where he continued to work as a research associate until 1987. He is a Principal Scientist/Product Manager sharing his time between R&D and product marketing at the Imaging Systems Division of Hewlett-Packard Company in Andover, MA. He is involved with exploring new materials and novel designs for the development of next generation ultrasonic transducers.

Haertling, Gene H.

Gene H. Haertling received his B.S. degree in ceramic engineering from the University of Missouri-Rolla in 1954. His M.S. and Ph.D. degrees, also in ceramic engineering, were earned from the University of Illinois in 1960 and 1951, respectively. From 1961 to 1973 he held staff and managerial positions at Sandia National Laboratories. During this time he developed the first transparent ferroelectric ceramics, the PLZT (lead lanthanum zirconate titanate) materials. From 1974 to 1987 he was Vice-President of the technical staff and manager of the ceramic research group at Motorola, Inc., Albuquerque, NM. In 1989 he joined the faculty of Clemson University and developed a special process for producing ultra-high-displacement actuators from piezoelectric ceramic materials. He is now Professor Emeritus at Clemson University and a member of the National Academy of Engineering. His technical interests have included electronic properties of ceramics, chemical processing, thin/thick films, ferroelectrics, piezoelectrics, electrooptics, and microwave ceramics.

Hastings, Ian J.

Ian J. Hastings has a B.Sc. (Hons) and a Ph.D., both in metallurgical science, from the University of Queensland, Australia. He is Director, Strategic Initiatives, for Atomic Energy of Canada, Limited, with a major responsibility for managing the strategic aspects of AECL's business. His primary research interests have been ceramic nuclear fuels, including fuel cycles, fabrication, irradiation behavior, and computer codes describing in-reactor performance; he has extended this interest to fusion ceramics. Hastings is the author of more than 220 technical publications, has edited three books on nuclear fuel, and coedited three on lithium ceramics for fusion applications. He is a Fellow and past Trustee of The American Ceramic Society, past Chair of the Nuclear Division of the Society, and sat on the Editorial Advisory Board of the Society. He is also a Fellow of the Australasian Institute of Mining and Metallurgy and is a member of the Editorial Advisory Board of the scientific journal *Nuclear Technology.*

Hench, Larry L.

Larry L. Hench received his B.S. in 1961 and his Ph.D. in 1964, both in ceramic engineering from The Ohio State University. He joined the faculty of the University of Florida in 1964 and remained there until 1996, when he accepted a University of London Chair at the Imperial College of Science, Technology, and Medicine. His interests include glass and biological materials.

Heuer, Arthur. H.

Arthur H. Heuer received his B.S. in chemistry from City College of New York, a Ph.D. in applied science from The University of Leeds, and a D.Sc. in physical ceramics from The University of Leeds in 1956, 1965, and 1977, respectively. He joined the Department of Materials Science and Engineering at Case Western Reserve University in 1967 as an Assistant Professor and is currently the Kyocera Professor of Ceramics. He was elected to the National Academy of Engineering in 1990 and was made an External Member of the Max Planck Institute for Materials Science, Stuttgart, Germany, in 1991. Professor Heuer is world-renowned for his research accomplishments on phase transformations in ceramics and intermetallics, transmission electron microscopy of defects in materials, high-resolution electron microscopy studies of interfaces in advanced structural composites, dislocations in ceramics, biomimetic processing of ceramics, MEMS, and rapid prototyping of engineering materials. He has published more than 380 articles in refereed journals and served as Editor of the *Journal of the American Ceramic Society* from 1988 to 1990. He was made a Distinguished Life Member of the American Ceramic Society in 1996.

Hinton, Jonathan W.

Jonathan W. Hinton obtained his B.S., M.S., and Ph.D. in ceramic engineering at The Ohio State University. Hinton held various positions at Champion Spark Plug Company and later served as Vice

President/General Manager, Structural Ceramics Division, Carborundum Company, a unit of British Petroleum. He is currently Senior Vice President, Lanxide Technology Company.

Hofmann, Doug

Doug Hofmann received both his B.S. and M.S. in geology and mineralogy from The Ohio State University. He is a research scientist at Owens Corning, where he has worked since 1979, first in the area of glass batch and raw materials and then in glass composition development. His major areas of interest are glass composition, glass physical properties, and computer programming.

Holowczak, John

John Holowczak received B.S. and M.S. degrees in ceramic engineering from Rutgers University in 1986 and 1989. He was employed by Norton/TRW Ceramics, where his efforts focused on development of complex-shaped fabrication for silicon nitride and silicon carbide turbine engine components. He is now a Research Engineer at United Technologies Research Center (UTRC), Advanced Ceramics and Coatings Group. His responsibilities include applications engineering and government program management for a wide variety of efforts examining the application of technical, structural, and composite ceramics and coatings in United Technologies' aerospace and commercial/industrial products.

Hyatt, Edmond P.

Edmond P. Hyatt received B.S. (1949), M.S. (1950), and Ph.D. (1956) degrees in ceramic engineering. He founded Electro-Ceramics, Inc., and later was employed by Centralab Division of Globe-Union, Inc. He is also the founder of EPH Engineering Associates, Inc. He consults on technical ceramics. His interests include casting and roll compaction of alumina ceramics and other tapes.

Ings, John B.

John B. Ings received his B.S., M.S., and Ph.D. from Rutgers University during the 1970s. For the past 12 years he has worked in the ferrite industry. His experience includes the growth of single-crystal ferrites at Litton Industries, microwave frequency ferrites at Trans-Tech, and megahertz frequency ferrites at Ceramic Magnetics. His principal interests are in the miniaturization and integration of ferrite components with computer chips and the casting of very large soft ferrite magnets.

Iwase, Makoto

Makoto Iwase received a Master of Industrial Physical Chemisry degree from Tokyo Metropolitan University in 1975. He is Director, Production Technology Development Center, Technology Development Division, Fabricated Glass General Division, Asahi Glass Company, Ltd. His technical interest is the development of advanced production technology for automotive glass.

Johnson, David W., Jr.

David W. Johnson Jr. received his B.S. in 1964 and his Ph.D. in 1968 from The Pennsylvania State University. He was a member of the technical staff at Bell Telephone Laboratories from 1968 to 1983, Supervisor of Advanced Ceramic Processing from 1983 to 1988, and Head of the Metallurgy and Ceramics Research Department from 1988 to the present. He has been an Adjunct Professor of materials science at the Stevens Institute of Technology since 1982. His research activities include fabrication and processing of spinel ferrites, synthesis and fabrication of high-T_c oxide superconductors, emphasizing the processing of ceramic shapes of potential technological interest; and sol–gel processing of glass and ceramics, emphasizing colloidal sols of fumed silica used to fabricate large pieces of transparent high-silica glass. Other interests include thin-film dielectrics prepared by sputtering and CVD, preparation of sodium-ion conductor materials, preparation of oxide catalysts, and characterization of ceramic powders. He is a member of the National Academy of Engineering. He received the Fulrath Award, the Ross Coffin Purdy Award, and the John Jeppson Award from The American Ceramic Society.

Johnson, James R.

James R. Johnson began his interest in ceramics very early throwing pots on a wheel. He became more interested in the technical side and obtained his university degrees at The Ohio State University. His career has included teaching, pioneering work on nuclear ceramics at Oak Ridge National Laboratory, and industrial R&D at 3M Company, where he was Laboratory Director for Physical Sciences and 3M Executive Scientist. His interests in science and technology have been very broad, but with emphasis on the development and use of many advanced materials and devices that have emerged over the past half century. Current interests include the biosciences, biomaterials, energy processes, polymers, and still throwing pots once in a while.

Kear, Bernard H.

Bernard H. Kear is a State of New Jersey Professor in the Department of Ceramic and Materials Engineering, Rutgers University. Kear received his B.Sc. and Ph.D. in physical and theoretical metallurgy, and his D.Sc. in materials science and engineering from Birmingham University, U.K. For more than 35 years, his research interests have been centered on the synthesis, processing, structure, and properties of inorganic solids, for a broad range of structural applications. Kear's current research is concerned with chemical processing of nanophase metals, ceramics, cermets, and composites, starting from aqueous solution or metalorganic precursors. Primary objectives of the research are to develop scalable processes for the production of nanostructured powders, thin films and multilayered structures, diffusion and overlay coatings, particle-dispersed and fiber-reinforced composites, and net-shaped bulk materials. Previous work, 1963–1981, at Pratt & Whitney, addressed fundamental aspects of dislocation interactions, phase transformations, and solidification behavior in nickel-based superalloys. This work contributed to the successful development of directional solidification of single-crystal turbine blades, and rapid solidification powder atomization and laser surface treatments. From 1981 to 1986, at Exxon, his research activities were focused on developing methods for CVD surface passivation treatments and for catalytic growth of carbon whiskers from hydrocarbon precursors. Kear has published more than 220 technical papers, edited nine books, and has been granted 35 patents. He was made a member of the National Academy of Engineering in 1979, was Chairman of NMAB from 1986 to 1989, and is Co-editor of the journal *Nanostructured Materials* (1992–present).

Keig, Tony

Tony Keig received his B.Sc. in physics from Liverpool University, England, in 1959 and his Sc.D. in materials science from the Massachusetts Institute of Technology in 1966. From 1959 to 1962 he was at the U.K. Atomic Energy Research Establishment, Harwell, doing research on physical properties of ceramic nuclear fuels. From 1966 to 1976 he was with Crystal Products, Union Carbide Corporation, concerned with development of oxide single crystals for electrooptic, semiconductor, and jewelry applications. From 1976 to 1987 he was with the Electronics Division, Union Carbide Corporation, in new business development. The primary business was a new high-purity source of polycrystalline silicon for the semiconductor industry.

Kelly, Robert J.

Robert J. Kelly holds a B.A. in chemistry from the University of California, a D.D.S. from The Ohio State University, an M.S. in dental materials science from Marquette University, and a D.Med.Sc. in oral biology with emphasis in ceramic engineering and glass science from Harvard University. Kelly is a guest scientist in the Dental and Medical Materials Group at the National Institute of Standards and Technology with academic appointments at George Washington University and at the Naval Dental School, as Director of Dental Materials. His research focuses on the fracture mechanics and clinical behavior of brittle materials. He serves on the Council on Scientific Affairs of the American Dental Association, and on the Editorial Board of the *Journal of Dental Research*, is a member of The American Ceramic Society, and is on active duty in the U.S. Navy with a limited practice in fixed prosthodontics.

Krug, Gene

Gene Krug received his Ph.D. in ceramic engineering from Rutgers University in 1991. He has been employed at Carpenter EPG Certech since then. His professional interests include ceramic injection molding, powder processing, and microstructure control.

Kugimiya, Koichi

Koichi Kugimiya graduated from Kyoto University and has both M.S. and Ph.D. degrees from the University of Texas. He also has a Dr. in Eng. from Tokyo Institute of Technology. He joined the

Matsushita Electric Company, Ltd., and has worked on magnetic materials, sensor materials, and other areas. He is now Executive Engineer, Corporate Research Division.

Kulwicki, Bernard M.

Bernard M. Kulwicki is a Distinguished Member of the Technical Staff, Texas Instruments, M&C Group, Attleboro, MA. He was educated at the University of Detroit (B.Ch.E., 1958) and the University of Michigan (Ph.D., 1963) and was Exchange Visitor at the Institute of Solid State Physics in Prague (1963–1964). He has been with TI since 1964. He is a Fellow of The American Ceramic Society and has served as an Associate Editor of the *Journal*. He is a contributor to the IEEE International Symposia on Applications of Ferroelectrics.

Lamicq, Pierre J.

Pierre J. Lamicq is in charge of Research and Technology Management at SEP, now a division of SNECMA. He graduated from the Ecole Polytechnique, Paris, in 1962. Since 1963, he has run rocket propulsion and High Performance Composites Research and Development activities in the SNECMA group, and mainly in Société Européenne de Propulsion (France). Materials and materials associations in complex systems have been an important part of his professional life: metals, catalyst supports, glass, Kevlar, carbon-reinforced composites, silica and carbon-reinforced phenolics, elastomers, carbon–carbon, and ceramic-matrix composites. Besides his industrial activities, Lamicq is the SEP delegate in LCTS. He is an associate member of the Conseil des Applications de l'Académie des Sciences (CADAS). He has coedited one book, published many articles, and holds two patents.

Lehman, Richard

Richard Lehman received his B.S. (1972), M.S. (1975), and Ph.D. (1976) degrees in ceramic engineering from Rutgers University. Prior to joining the Rutgers faculty in 1982, Lehman gained eight years of experience at Johns-Manville and FMC Corporations where he worked in the glass, fiberglass, and industrial chemical fields. In 1982, Lehman joined the faculty at Rutgers, where he is now Professor of Ceramic and Glass Science Engineering. He conducts basic and applied research in new and traditional glass and ceramic materials. He has particular interest in conventional glass-melting processes, convective flows in glass tanks, optical glasses, and fiber optics and sensors. He also is known for his work on the chemical durability of glass and ceramics with a current focus on the behavior of lead-containing glass and ceramic foodware surfaces. Lehman is the Assistant Director of the Fiber Optics Materials Research Program at Rutgers. In addition to research activities, Lehman is active in graduate and undergraduate instruction, and he participates in external consulting activities with local, national, and international industries and organizations. Lehman has more than 70 publications and 16 patents on the processing and properties of glass and ceramic materials and he is a Fellow of The American Ceramic Society.

Levinson, Lionel M.

Lionel M. Levinson, received his B.S. and M.S. in physics and mathematics from the University of the Witwatersrand, South Africa, in 1965 and 1966. His doctoral studies in solid-state physics were undertaken at the Weizmann Institute of Science, Rehovot, Israel. Levinson joined the research staff of GE Corporate Research and Development in 1970 and is now Manager, Electronic and Optical Materials Program, GE Corporate Research and Development. His research has been primarily focused on the development of novel composite materials and electronic ceramics. In 1980 Levinson was corecipient of an IR-100 award for the development of a varistor-switched liquid crystal display. Levinson has written or edited four books, approximately 85 technical papers and articles pertaining to his fields of work, and has been awarded 36 U.S. patents.

Li, Chien-Wei

Chien-Wei Li received his B.S. in materials science and engineering from Tsing Hua University, Taiwan, in 1979, and his Ph.D. in ceramics from the Massachusetts Institute of Technology in 1987. Currently, he is a Senior Principal Scientist in the Research and Technology Center of AlliedSignal, Inc., focusing mainly on the development and implementation of high-temperature ceramics for advanced engine applications.

Logothetis, Eleftherios M.

Eleftherios M. Logothetis received his B.S. in physics from the University of Athens, Greece, in 1959, and his M.S. and Ph.D. degrees in solid-state physics from Cornell University, in 1965 and 1967, respectively. He has been with Ford Motor Company since 1967, and he is currently a Senior Staff Scientist in the Ford Research Laboratory. He is a Fellow of the American Physical Society. His main research has been in the preparation and the electrical and optical properties of materials, solid-state devices, in particular, chemical sensors.

Lukasiewicz, Stanley

Stanley Lukasiewicz received a Ph.D. in ceramics from Alfred University in 1984. He worked at Stackpole Carbon Company and is now at Texas Instruments. His areas of technical interest are powder processing, particularly powder granulation and compaction, PTCR thermistors, LTCC–metal cofire systems, thick-film paste systems, magnetic ceramics, and ceramic capacitive sensing elements.

MacChesney, John B.

John B. MacChesney received his B.S. and Ph.D. degrees from Bowdoin College, Brunswick, ME, and The Pennsylvania State University, respectively. Since graduating from Penn State in 1959, he has worked for Bell Laboratories on materials-related problems and since 1974 on optical materials. He has received many awards, and he is a member of The American Ceramic Society, the Institute of Electrical and Electronic Engineers, and the National Academy of Engineering.

MacZura, George

George MacZura earned his B.S. in ceramic engineering from the University of Missouri-Rolla in 1952 and was awarded the Professional Degree of Ceramic Engineer in 1972. After 44 plus years of service with ALCOA, MacZura retired on January 1, 1997, from his position of Refractory Market Development Manager–International for the North American Industrial Chemicals Division in Pittsburgh, PA. He is the coinventor of Alcoa's calcium aluminate cements, reactive aluminas, and improved tabular aluminas, which were originated, developed, and commercialized during his 34 years with Alcoa Laboratories. These products are used by almost all heat processing and ceramic industries throughout the world. MacZura holds seven patents, has published/presented more than 40 technical papers, and has authored/coauthored about 70 internal reports.

Maguire, Edward A.

Edward A. Maguire received a B.S. in ceramic engineering from Alfred University in 1957. He has over 45 years of experience with ceramics for electronics and other advanced technology applications. He has worked for International Resistance Company, Philadelphia, PA, and Raytheon Electronic Systems in Lexington, MA. His interests have been in ferrites, ferroelectrics, piezoelectrics, and transparent ceramics.

Marra, Robert A.

Robert A. Marra received at B.S. in ceramic science and a B.A. in mathematics in 1978 from Alfred University. He completed his Ph.D. in ceramic science from the Massachusetts Institute of Technology in 1982. Since his graduation from MIT, he has been employed by Alcoa. Marra was a member of the technical staff at the Alcoa Technical Center from 1982 to 1997. During that period, he served in various R&D roles supporting the ceramic and refractory needs of Alcoa business units ranging from smelting (primary metals), industrial chemicals, ingot plants, and electronic packaging as well as new ventures in cermets, metal- and ceramic-matrix composites, and advanced ceramics. From 1992 to 1997, he served as the Manager of the Ceramics Technology Center. In April 1997, Marra joined the Alcoa Industrial Chemicals Business Unit as Senior Application and Market Development Manager for refractory materials. In April 1998, Marra assumed his current position as the Manager of the North American Industrial Chemicals Application and Product Development Groups.

Maurer, Christopher

Christopher Maurer received his Ph.D. in ceramic engineering, with a minor in anthropology, from the University of Illinois, his M.S. in physics, with a minor in environmental engineering, from the University of Nebraska, and his B.S. in physics from the University of Illinois. After working with thermoluminescence and related dating methods for many years, he moved into environmental engineering and is presently working as an environmental engineer with the State of Washington's Department of Ecology. He is affiliated with the Design Division of The American Ceramic Society.

McCandlish, Larry E.

Larry E. McCandlish holds a B.S. from Case Western Reserve University and a Ph.D. in physical chemistry from the University of Washington. He has 21 years of experience developing a wide variety of materials, which find applications ranging from catalysts to cutting tools. He was employed for 10 years at Exxon Research and Engineering Company, where he specialized in making new catalytic materials. He joined the Department of Mechanics and Materials Science at Rutgers University in 1987 and, based on his work at Exxon and Rutgers, cofounded Nanodyne Incorporated, where he was Vice President of Research between 1992 and 1996. He is currently a member of Nanodyne's Board of Directors. In 1996 he cofounded Ceramaré Corporation to develop and commercialize hydrothermal processes for the production of electroceramic materials. He now serves as president of Ceramaré. He has authored more than 40 technical publications, edited one book, and has been granted 19 patents.

McConnell, Robert

Robert McConnell received a B.S. in engineering from the University of Alabama. He has been employed at Alcoa Industrial Chemicals since February 1969, where he is now Applications and Market Development Manager for Refractories. Primary technical interest areas are high-purity monolithic refractories and aluminum oxide-based ceramics.

Mistler, Richard E.

Richard E. Mistler received a B.S. in ceramic engineering from the New York State College of Ceramics at Alfred University, an M.S. in metallurgical engineering from Rensselaer Polytechnic Institute, and an Sc.D. in ceramics from the Massachusetts Institute of Technology. He is owner and President of Richard E. Mistler, Inc., a consulting and contract R&D firm involved in technical ceramics. Mistler held previous positions at American Oil and Supply International, where he was Vice President of Research, and Western Electric Company, where he was a Research Leader in Ceramic and Chemical Processes. He is the author of more than 100 presentations and publications and is a Fellow of The American Ceramic Society. Mistler is currently collaborating with Eric Twiname on a book on tape casting that will be published by The American Ceramic Society.

Mulcahey, Brian D.

Brian D. Mulcahey holds an M.B.A. in high-technology marketing from the University of Chicago and a B.S. in electrical engineering from Massachusetts Institute of Technology. He has worked as a technical consultant for Booz Allen & Hamilton, a Business Planning Analyst at Ford Motor Company, and a Product Marketing Manager with GenRad. He is now a Senior Marketing Manager at ACX, where he leads the marketing efforts for ACX's Vibration and Motion Control Group. His responsibilities include business development, product management, market research, public relations, and marketing communications.

Naslain, Roger R.

Roger R. Naslain received his Ph.D. in materials science from the University of Bordeaux in 1967. In 1968 he was a visiting researcher at the R&D Center of General Electric Company in Schenectady, NY. From 1969 to 1982 he was head of a research group on metal-matrix composites at CNRS Laboratory for Solid State Chemistry. From 1983 to 1987 he was Director of the Institute for Composite Materials. Since 1988 he has been director of LCTS, a joint institution involving CNRS, the University of Bordeaux, and a company from the aerospace industry, SEP Division of SNECMA. During the past 10 years, Naslain has conducted research on ceramic-matrix composites (CMCs) in terms of processing (CVI process), interface design (multilayer concept), and the effect of the environment of non-oxide CMCs. He has edited or coedited six books, published more than 200 articles, and has 15 patents.

Nelson, Michael A.

Michael A. Nelson received a B.S. in ceramic engineering from Iowa State University in 1965. He was employed by PPG Industries from 1965 to 1966 and by Corhart Refractories from 1966 to the present. He has worked in process engineering, research and development, technical services, glass refractory sales, product management, and is currently in applications engineering, primarily in the area of refractories for the glass industry.

Newnham, Robert E.

Robert E. Newnham received a B.D. in mathematics from Hartwick College in 1950, a M.S. in physics from Colorado State University in 1952, and two Ph.D.s, the first in physics and mineralogy from The Pennsylvania State University in 1956, and a second in crystallography from Cambridge University in 1960. After eight years at the Laboratory of Insulation Research at Massachusetts Institute of Technology, he joined the faculty at Penn State, where he is Alcoa Professor Emeritus of Solid State Science and Associate Director of the Materials Research Laboratory. His research interests are in crystal physics, crystal chemistry, electroceramics, and functional composite materials for use as transducers, sensors, and actuators. Newnham in a member of the National Academy of Engineering and a Distinguished Life Member of The American Ceramic Society.

Niesz, Dale E.

Dale E. Niesz received his B.S. (cum laude), M.S., and Ph.D. in ceramic engineering from The Ohio State University. He is Professor and Chair of the Ceramics and Materials Engineering Department, Rutgers University, where he has worked since 1987. Since 1988 he has also served as Director of the Center for Ceramic Research at Rutgers. Prior to joining the Rutgers faculty, he worked at Battelle's Columbus Laboratories for 22 years, including eight years as Manager of the Materials Department. His interests are in ceramic powder processing, especially the relation of powder characteristics, processing parameters, and chemistry to microstructure development in ceramic materials.

Nishikori, Tsuneharu

Tsuneharu Nishikori received a Bachelor of Inorganic Chemistry in ceramics from the Tokyo Institute of Technology in 1961. He is Executive Vice-President of the Asahi Glass Company, Ltd. His technical interest is the development of innovative glass processing and production and enhancing the capability of glass for tomorrow.

O'Bryan, Henry M.

Henry M. O'Bryan received his B.S. in chemical engineering from the University of Notre Dame and his M.S. and Ph.D. in engineering materials from the University of Michigan. Since 1963 he has been a member of the technical staff at Bell Laboratories, AT&T/Lucent Technologies. His technical interests cover many aspects of electronic ceramics, including ferrites, ferroelectrics, microwave dielectrics, microstructure, processing, and, more recently, thin-film dielectrics and electrodes.

Owens, James S.

James S. Owens received his B.S. in physics from University of Chattanooga (now University of Tennessee at Chattanooga) in 1926, his M.S. in physics from the University of Michigan in 1930, and his Ph.D. from the University of Michigan in 1930. From 1933 to 1943 he was engaged in spectroscopic research at Dow Chemical Company. He was with the NDRC, OSRD from 1943-1945, and was a professor at The Ohio State University from 1945 to 1951. He was with Champion Spark Plug Company from 1951 until his retirement in 1973. He was Vice President and General Manager of the Ceramic Division at Champion Spark Plug Company.

Palmquist, Ron

Ron Palmquist obtained a B.S. from Princeton University and a Ph.D. in 1970 from the University of Wisconsin, both in mechanical engineering. After graduating, he joined Corning Glass Works in the Furnace Development Department and participated in the development of electric furnaces. He is now a Senior Engineering Associate in the Glass and Glass–Ceramics Department.

Pastor, Henri

Henri Pastor is an engineer of Ecole Nationale Superieure d'Electrochimie et d'Electrometallurgie de Grenoble and has a Ph.D. from the University of Grenoble. He was a research engineer with Ugine Carbone Company, Managing Director of R&D Laboratories at Ugicarb-Morgon Company, and Chairman and Managing Director of CERMEP (European Research Center on Powder Metallurgy). His most recent activities were on tungsten carbide, cemented carbides, and cermets.

Pickering, Michael A.

Michael A. Pickering received a B.A. in chemistry from Rhode Island College in 1976 and Ph.D. in physical chemistry from Brown University in 1982. Pickering worked as a Research Associate in the chemistry department at the University of Rhode Island for one year and began working at Morton Advanced Materials (formally CVD, Inc.) in 1983 as a Principal Research Scientist. Since joining Morton, he has been the lead scientist and Program Manager on numerous in-house, government, and privately supported programs including development of infrared (IR) optical fibers, IR gradient index material, electroluminescent materials for flat panel displays, IR rugate filter, electromagnetic shielding for IR transmissive materials, the chemical vapor deposition (CVD) process to increase product yield and quality, and large-scale, high-temperature ceramic CVD production furnaces. The most recent accomplishment is the development and large-scale production of CVD SiC. In July 1992 Pickering became the Market Manager for Morton's Advanced Ceramics business with responsibility for worldwide sales and marketing of CVD Silicon Carbide® products. He currently holds the position of Technical Manager and is responsible for all research and development activities within Morton Advanced Materials. Pickering was the recipient of the Heroes of Chemistry Award by the American Chemical Society in 1996 and the Engineering Excellence Award in 1991 by the Optical Society of America.

Pinckney, Linda R.

Linda R. Pinckney received her B.S. in earth and space sciences from the State University of New York at Stony Brook in 1976 and her A.M. and Ph.D. in mineralogy from Harvard University in 1978 and 1982. She joined Corning in 1982, where she is a Senior Research Associate in the Science and Technology Division. Her current research interests include the thermal and optical properties of transparent glass-ceramics and new applications for them. She holds 12 U.S. patents in glass and glass-ceramic materials, and is a member of The American Ceramic Society, the Materials Research Society, and the Mineralogical Society of America.

Pope, Edward J. A.

Edward J. A. Pope received his Ph.D. in materials science and engineering in 1989, his M.S. in 1985, and B.S. in 1983, all from the UCLA School of Engineering and Applied Science. He has published more than 57 peer-reviewed technical papers, chaired or cochaired four international research meetings, edited or coedited five monographs and proceedings, received seven patents, with several more pending, and received numerous international awards, including the W. A. Weyl

International Glass Science Award, and the D. R. Ulrich Award for "outstanding research in the field of sol–gel science and technology." In 1989, Pope founded MATECH, a private research and development laboratory that specializes in contract research for corporate clients in the areas of optical and biomedical materials. In 1997, Pope launched a new spin-off company, Solgene Therapeutics, LLC, to commercialize technology for biotech drug delivery and cell therapy. In 1998, Pope has launched FPD Technologies, to commercialize new flat panel display technology, and Global Strategic Materials, in the area of preceramic polymer technology for ceramic-matrix composites.

Prindle, William R.

William R. Prindle received B.S. and M.S. in physical metallurgy from the University of California at Berkeley and received Sc.D. in ceramics from Massachusetts Institute of Technology. In 1954 he joined the Hazel-Atlas Glass Division, Continental Can Company, and became General Manager of R&D before leaving in 1962 to become Manager of Materials Research, American Optical Company. In 1966 he joined Ferro Corporation in Cleveland, where he became Vice President–Research. In 1971 he returned to American Optical Corporation where he was Vice President and Director of Research until 1976. From 1976 to 1980 he was in Washington, DC, as Executive Director of the National Materials Advisory Board, a unit of the National Research Council of the National Academy of Sciences. He joined Corning Incorporated at the end of 1980 and retired in 1992 as Division Vice President and Associate Director, Technology Group.

Prochazka, Svante

Svante Prochazka, a native of the former Czechoslovakia, retired from General Electric after serving 24 years as ceramist with Corporate R&D. He holds an engineering chemistry degree and Ph.D. from the Institute of Chemical Technology, Prague, and an S.M. in ceramics from Massachusetts Institute of Technology. Early in his career he worked as a research fellow, research chemist, and manager in the spark plug and electronic industries. At GE he worked mostly in synthesis and processing of structural, optical, and electronic materials. He is a consultant to GE at present.

Racher, Ray

Ray Racher has a B.S. in ceramic science and technology from The University of Sheffield, England, and has worked in various technical/commercial roles in the ceramics and refractories industries since 1977. He is employed at Alcoa, Inc.

Rankin, D. Thomas

D. Thomas Rankin received his B.S. in 1963 and his Ph.D. in 1967, both in ceramics from Rutgers University. Since 1967 he has worked at the Savannah River Site, except for military duty during 1968–70 at the U.S. Army Materials and Mechanics Research Center. His research interests have concentrated on the fabrication of nuclear fuels and immobilization forms with emphasis on the technical basis for each processing step. Currently, Rankin is an R&D Team Leader and Group Manager at the Savannah River Technology Center. He also is a Fellow of The American Ceramic Society and served on the Board of Trustees.

Rapp, Charles

Charles Rapp received both his B.S. and Ph.D. in chemistry from the University of Toledo. He worked at Owens Illinois for 17 years and is now a research scientist at Owens Corning, where he has worked for 19 years. His primary research areas and interests include fiberglass, glass lasers, optical and technical glass, heavy-metal fluoride glasses, and mineral wool.

Rhodes, William

William Rhodes earned his B.S. in ceramic engineering from Alfred University in 1957. During 1960–1961 he pursued graduate studies in metallurgy at Rensselaer Polytechnic Institute. He received his Sc.D. in ceramic science from Massachusetts Institute of Technology in 1965. He is Manager of Lighting Materials Research Department for OSRAM SYLVANIA, Inc., in Beverly, MA. His research specialty is sintering oxides to optical transparency and studying their physical properties. Rhodes served as President of The American Ceramic Society in 1988–1989 and is a member of the Academy of Ceramics.

Rokhvarger, Anatoly E.

Anatoly E. Rokhvarger received Ph.D. and D.Sc. in ceramic engineering and materials science from Moscow and Leningrad Chemical–Technological Universities. Working at R&D Ceramic Industry Analytical Center and Laboratories in Moscow, he contributed development of nine technological systems and six ceramic products, including conveyer technology for thin-walled ceramics, production of porous materials in fluidized bed furnace, and production of continuous BN fibers and wear-resistant magnetic granules. He published 144 articles, three books, and one textbook. Since 1992 he has been Research Professor, Polytechnic University, Brooklyn, NY. Since 1997, as a Vice-President of R&D of Nucon Systems, Inc., he has been running his project, "Large-Size and Thick-Walled Ceramic Containers for Nuclear and Hazardous Waste," working at the Center of Ceramic Research, Rutgers University, NJ.

Roth, Robert

Robert Roth received a B.S. in electrical engineering from the University of Florida and did graduate studies in business at the University of California, Berkeley. He is the marketing manager for Marathon Ceramics, a division of Mountain Safety Research (MSR), Seattle. His experience includes marketing management and introduction of new technologies for Tektronix, Fluke, and Philips NV, as well as the development of new market opportunities for MSR. Roth is named on a patent concerning electronic instrumentation and digital signal processing.

Roy, Rustum

Rustum Roy received a Cambridge School Certificate from St. Paul's School in Darjeeling, India, in 1939. He received a B.Sc. in 1942 and an M.Sc. in 1944 in physical chemistry from Patna University. He holds a Ph.D. in ceramics, 1948, from The Pennsylvania State University. He has been employed by The Pennsylvania State University since 1950, where he is Evan Pugh Professor of the Solid State, Professor of Science, Technology, and Society, and Professor of Geochemistry. His technical interests include new-materials preparation and characterization. Areas of specialization are crystal chemistry, synthesis, stability, phase equilibria and crystal growth in nonmetallic systems, ultra-high-pressure reactions in solids, radioactive waste forms, nanocomposites, zero-expansion ceramics, and diamond films. He is a principal architect of sol–gel and hydrothermal processes and is active in microwave processing. He is active in science, technology, and human values; science and public policy; technology and religion; and technological literacy for all citizens.

Rubin, Jack A.

Jack A. Rubin holds a B.S. in geochemistry and attended graduate school at the University of Califonia. He cofounded Ceradyne, Inc., where he served as Chair and Technical Director. Later he was Vice-President and Technical Director of Kyocera. Subsequently, he served as Staff Scientist (Ceramics) for Raychem Corporation. Rubin then was the Vice-President and Technical Director of Microelectronic Packaging, Inc. He has been a very successful consulting ceramic engineer/scientist since 1989. He specializes in CMC and MMC composites, structural ceramics, ceramic microelectronic packaging, and hot pressing. Rubin is a member and past-president (San Diego Chapters) of The American Ceramic Society, IMPS, and ISHIM (now known as IMAPS). He also has been a member of ACS, ASTM, ASEF, and MRS. He has authored or coauthored many publications and patents.

Ruh, Edwin

Edwin Ruh received his B.S. and M.S. in ceramic engineering and his Ph.D. in ceramics from Rutgers University. He was Vice-President of Research at Vesuvius and Director of Research at Harbison–Walker Refractories. Subsequently he spend eight years with the Department of Metallurgical Engineering and Materials Science at Carnegie-Mellon University, at which time he was editor of *Metallurgical Transactions.* He spent the next 10 years at Rutgers University, Department of Ceramics, where he was a research professor and helped with the planning and construction of the McLaren Center for Ceramic Research and the Fiber-Optics Buildings. He is a consultant on refractories.

Sakamoto, Terry

Terry Sakamoto received B.S. and M.S. in chemistry from Kagoshima University. He is the Automotive Components Development

Manager at Kyocera Corporation, Kokubu, Japan, where he is responsible for the development of silicon nitride components for automotive and industrial applications. While at Kyocera, he has been responsible for developing prototype and production manufacturing methods for silicon nitride components. Processes developed by Sakamoto are currently used for numerous components utilized in gas turbine, automotive, and diesel engines.

Sato, Yasuo

Yasuo Sato received B.S. and M.S. in electrical engineering from Shinshu University, Ngano, Japan, in 1970 and 1972, respectively, and a Ph.D. from Shinshu University in 1994. His Ph.D. work dealt with a new float process by using a linear induction motor. He is the Director of Technology and Development Division of CRT Glass Bulbs General Division, Asahi Glass Company, Ltd. His principal interest is in development of innovative CRT glass production processes and new advanced products. He is a member of the Institute of Electrical Engineers of Japan and the Japan Society of Mechanical Engineers.

Shackelford, James F.

James F. Shackelford has B.S. and M.S. in ceramic engineering from the University of Washington and a Ph.D. degree in materials science and engineering from the University of California, Berkeley. He is currently a professor in the Department of Chemical Engineering and Materials Science and the Associate Dean for Undergraduate Studies in the College of Engineering at the University of California, Davis. A member of The American Ceramic Society, he was named a Fellow of the Society in 1992 and received the Outstanding Educator Award in 1996.

Sheldon, David A.

David A. Sheldon received B.S. and M.S. in ceramic engineering from Alfred University with respective efforts in the oxidation study of SiC refractories and fabrication and characterization of ceramic fiber/ceramic matrix composites. He is employed as a research and development scientist in the Abrasives Branch of Saint-Gobain Corporation. He has held various research and supervisory positions for Norton and Saint-Gobain over a period of 24 years (in Worcester, MA), with interests in product development, process problem solving, and relationships between physical properties and product performance.

Singh, Mrityunjay

Mrityunjay Singh is Chief Scientist, FDC-NYMA, Inc., NASA Lewis Research Center, Cleveland, OH. After earning a Ph.D. in metallurgical engineering from Banaras Hindu University, Varanasi, India, he held various research assignments at Rensselaer Polytechnic Institute, Troy, NY, and Case Western Reserve University, Cleveland, OH. His research interests include processing and characterization of advanced ceramics and fiber-reinforced ceramic composites, ceramic joining, crystal growth of optoelectronic materials, and biomaterials. The author or coauthor of more than 115 publications and one edited book, he is a Fellow of ASM International and The American Ceramic Society. He is recipient of numerous awards including Jacquet Lucas Award from IMS-ASM International, R&D 100 Award, ASM International-IIM Visiting Lectureship, NYMA Innovator of the Year Awards, Enterprise Development Institute (EDI) Innovation Award, and Federal Laboratory Consortium (FLC) Award for Excellence in Technology Transfer.

Skaggs, S. R.

S. R. Skaggs received his B.S. in mechanical engineering in 1958, his M.S. in nuclear engineering in 1967, and his Ph.D. in materials science in 1972. His principal employment was at the Los Alamos National Laboratory, where he was Armor Manager from 1985 until his retirement in 1993. His principal interests include phase equilibria, ceramic–metal joining, high-temperature materials, radiation damage, energy conversion systems, and ceramic armor.

Skandan, Ganesh

Ganesh Skandan received his B.S. in metallurgical engineering from the Indian Institute of Technology, Bombay, India, in 1990 and his M.S. in materials science from Rutgers—The State University of New Jersey in 1992. He received his Ph.D. degree in 1995. Skandan is a cofounder of Nanopowder Enterprises, Inc., and is Vice-President—Research and Development. He also is a cofounder and Vice-President of the holding company, SMT. His work has focused on the area of processing ultra-fine-grained materials, particularly powder synthesis and sintering to dense structures with grain sizes in the nanoscale range. He was coinventor of the modified patented version of the chemical vapor condensation (CVC) process.

Sowman, Harold G.

Harold G. Sowman received his B.S. (1948), M.S. (1949), and Ph.D. (1951) in ceramic engineering at the University of Illinois. Employed by Titanium Alloy (TAM) Division of National Lead Company, 1951–1952, General Electric Knolls Atomic Power Laboratory, 1952–1957, and 3M Company, 1957–1987. Career emphasis to 1965 was on refractory ceramics and powder metallurgy, cermet, and ceramic nuclear fuel materials. The major portion of his career was at the 3M Company, where he initiated and conducted—for more than 25 years—a program on the refinement of chemical ceramic or sol–gel technology for the development and manufacture of new ceramics. Commercial products resulting from this program, and believed to be the first successful ones of their type in industry from sol–gel or chemical ceramic technology, include spools of continuous ceramic filament strands, reflective ceramic (not glass) beads for highway marking, and abrasive granules.

Spangler, Ronald L., Jr.

Ronald L. Spangler Jr. holds a B.S. and an M.S. from Massachusetts Institute of Technology. He received his Ph.D. in control and estimation from the Department of Aeronautics and Astronautics at MIT. He leads the engineering efforts for ACX's Vibration and Motion Control Group. His technical responsibilities include development of control systems, based on smart-material (piezoelectric, electrostrictive, etc.) actuators and sensors, for applications involving structural vibration suppression, isolation, pointing, and tracking. Spangler has led ACX programs involving the application of smart-materials technology to sporting goods, fighter jets, machine tools, helicopters, and precision optical metering structures.

Strasser, Thomas A.

Thomas A. Strasser received a B.S. from Alfred University in 1987 and M.S. (1989) and Ph.D. (1993) in materials science and engineering from Cornell University. During his graduate studies he also worked at Eastman Kodak Corporate Research Laboratories. He has been a researcher at Bell Laboratories in Murray Hill since 1993, where he has worked primarily on silica waveguide and fiber devices fabricated with photo-induced index changes. Presently he is a Technical Manager of the Fiber Gratings and Devices Group in the Optical Fiber Research Department. His group researches the design of optical fiber, fiber grating, and other novel fiber devices for future application in lightwave communication systems. He has contributed more than 30 papers on gratings and guided-wave devices and as well as authoring 10 patents.

Sturgis, David H.

David H. Sturgis is Manager of the Non-Metals Section of the Materials and Technology Department at PCC Structurals, Inc., a Division of Precision Castparts Corporation, Portland, OR. He received his B.S., M.S., and Ph.D. in ceramic engineering, all from the University of Illinois in 1961, 1965, and 1968, respectively. Sturgis has been involved in material and process development programs in the investment casting industry for almost 30 years, primarily in aerospace and industrial gas turbine applications.

Taft, Karl

Karl Taft is a Ceramic Research Assistant at PCC Structurals, Inc., a Division of Precision Castparts Corporation, Portland, OR. He received his M.S. in 1995 in environmental science, specializing in environmental chemistry, from the Oregon Graduate Institute of Science and Technology, Portland, OR. His principal technical interests include completely combustible binders for slurries and environmentally friendly polymers.

Tajima, Kisuke

Kisuke Tajima received a Bachelor of Inorganic Chemistry from Waseda University in 1959. He is Corporate Technical Advisor at the Asahi Glass Company, Ltd. His interests are in development of new refractories and new glasses.

Thompson, Jim

Jim Thompson received his B.S. in ceramic engineering from Iowa State University in 1959. He worked for American Lava/3M from 1959 to 1980, where he had the technical responsibility for the tape-casting process from 1959 to 1969. He was the division Process & Industrial Manager until 1980. From 1980 to 1997 he was involved in

3M's total quality process and was the creator and head of 3M's Quality Management Services, an external consulting business, from 1958 until his retirement and the closing of that business on January 1, 1997. He is currently President of QMS of Minnesota, a continuation of the former 3M Quality Management Services business.

Thurnauer, Hans

Hans Thurnauer was born in 1908 in Nuernberg, Germany. He spent all his professional life in the pursuit of technical ceramics, from A(lumina) to Z(irconia). To this end he studied chemistry, ceramics, and engineering in Germany and the United States and earned Dipl. Ing., Dr. Ing. and M.Sc. He was employed by American Lava Corporation, Chattanooga, TN, 3M Company, St. Paul, MN, Coors Porcelain Company, Golden, CO, and IESC (International Exchange Service Corporations).

Tiegs, Terry

Terry Tiegs holds B.S. and M.S. in ceramic engineering from the University of Illinois and a Ph.D. in metallurgical engineering from the University of Tennessee. He is a senior development staff member in the Metals and Ceramics Division at the Oak Ridge National Laboratory (ORNL). Since joining ORNL in 1975, he has been involved in numerous research areas, including nuclear fuel behavior, ceramic–metal systems, microwave processing, and whisker-reinforced composites. Recently, his research focus has been on structural ceramics for advanced heat engines. Resulting from these efforts, he has received several awards, including Martin Marietta Energy Systems Inventor of the Year and the Federal Laboratories Consortium Excellence in Technology Transfer. In 1990 he was a recipient of the Purdy Award from The American Ceramic Society. Tiegs is the author or coauthor of more than 90 technical publications and holds 15 U.S. patents.

Tressler, Richard E.

Richard E. Tressler is Professor and Head of the Department of Materials Science and Engineering at The Pennsylvania State University, where he has been a faculty member since 1972. He received his B.S. and Ph.D. at Penn State with an intervening M.S. degree from Massachusetts Institute of Technology. He served four years as a U.S. Air Force officer. He is a Past President and a Fellow of The American Ceramic Society and also has received two of its national awards. He currently chairs the University Materials Council, the national organization of department heads of materials science and engineering departments. Tressler has authored more than 200 papers in refereed journals, proceedings, and books. He is the contributing editor of six books.

Twiname, Eric R.

Eric R. Twiname earned a B.S. in ceramic engineering from the New York State College of Ceramics at Alfred University. He is currently on an educational leave from Richard E. Mistler, Inc., while pursuing a Ph.D. in ceramics at The Pennsylvania State University. Mr. Twiname was Chief Engineer at Richard E. Mistler, Inc., and Operations Manager of the Tape Casting Warehouse for six years after a brief employment as a machine design engineer at Loomis Products Company. Mr. Twiname is currently collaborating with Richard Mistler on a book on tape casting that will be published by The American Ceramic Society.

Uram, Stuart

Stuart Uram obtained his S.B., S.M., and Sc.D. from the Massachusetts Institute of Technology. He was the founder and President of Certech, Inc. He is presently retired President of Certech, Inc.

Wachtman, John B.

John B. Wachtman received his B.S. and M.S. in physics from Carnegie Institute of Technology (now Carnegie-Mellon University) and his Ph.D. in physics from the University of Maryland. He worked from 1951 to 1983 at the National Bureau of Standards, becoming Director of the Center for Materials Science for the last five years. He then worked at Rutgers University from 1983 through 1995 as Sosman Professor of Ceramics and as Director of the Center of Ceramic Research from 1983 through 1988. He has been overall technical editor for The American Ceramic Society since 1988. He is the author of more than 160 papers and of the books *Characterization of Materials and Mechanical Properties of Ceramics*. He is a member of the National Academy of Engineering and the Academy of Ceramics. His interests are mechanical and electrical behavior of ceramics, characterization of materials, science and public policy, and the history of technology.

Walker, Kenneth

Kenneth Walker received a B.S. and Ph.D. in chemical engineering from Caltech and Stanford University in 1974 and 1979, respectively. He joined Bell Laboratories in 1979. His work has resulted in an improved understanding of optical-fiber fabrication, improved production processes, and new fiber designs. He is a Bell Laboratories Fellow and currently heads the Optical Fiber Research Department at Lucent Technologies, which includes responsibility for both transmission fiber and fiber amplifiers. Walker has been granted 35 U.S. patents.

Webb, Jayne M.

Jayne M. Webb graduated in 1993 from the University of Cincinnati College of Engineering, Cincinnati, OH, with a B.S. in materials engineering with a concentration in ceramics. She began working with the Unifrax Corporation (formerly Carborundum Company) in 1990 as a co-op student and gained experience in all facets of the company, including manufacturing, technical development, quality assurance, and marketing. In 1993 she was hired as a full-time application engineer and supported the furnacing group—supplying engineered furnace lining quotations. She now supports sales, marketing, and development for the converter product lines, including ceramic-fiber papers, boards, felts, and bulk fiber.

Wenckus, Joseph F.

Joseph F. Wenckus received his B.S. degree in chemical engineering from Northeastern University. He was a member of the professional staff at Airtron Division of Litton Industries, Microwave Associates, and Lincoln Laboratory. In 1963 he joined Arthur D. Little, Inc., where, for 14 years, he directed and participated in a broad range of research and development programs relating to crystal synthesis, technical audits, and market analyses for clients worldwide. As President of Intermat Corporation (an ADL Enterprise) he also was responsible for the development and marketing of the ADL high-pressure crystal-growing furnace systems. In 1976 Wenckus cofounded and was named President of Ceres Corporation to pursue the development of skull-melting technology and the manufacture of refractory oxide materials using this unique process. Under his direction, Ceres became the world's largest producer of single-crystal cubic-zirconia, which is used by the gem industry as a diamond simulant. He retired from Ceres Corporation in 1994 and formed Zirmat Corporation to provide a variety of single-crystal refractory oxides to the international high-tech community. Wenckus has lectured extensively in his areas of expertise—high-pressure crystal growth and skull-melting techniques—and holds 13 patents relating to crystal process technology and related equipment design. In 1992 Wenckus and Professor V. V. Osiko (Lebedev Physics Institute, Moscow) were jointly awarded the Laudise Prize by the International Organization of Crystal Growth for their development and commercialization of skull-melting technology for the large-scale production of single-crystal cubic-zirconia.

Wicks, George G.

George G. Wicks received a B.S. in engineering science and a S.M. in materials science from Florida State University. He received a M.S. in applied physics and engineering from Harvard University and a Ph.D. in metallurgy and materials science from Massachusetts Institute of Technology. He is a Senior Advisory Scientist at the Westinghouse Savannah River Site in Aiken, SC, and has been involved for more than 25 years in many areas of glass science, including vitrification of radioactive and hazardous wastes. Among his responsibilities was design and development of the first slurry feeding system for vitrification of SRS high-level radioactive waste. He also was codeveloper of the SRS kinetic leachability model used to describe nuclear glass behavior and principal investigator for *in-situ* testing programs involving burial of simulated U.S. nuclear glass compositions in geologic formations in four different countries. Dr. Wicks has published more than 150 papers, six invited book chapters, and coauthored 4 books. He is currently President of the National Institute of Ceramic Engineers and a Fellow of The American Ceramic Society. He also was past Chairman of the Nuclear Division of The American Ceramic Society, a member of the U.S. Materials Review Board, and is an adjunct Professor to Clemson University.

Wilcox, David L., Sr.

David L. Wilcox Sr. received his B.S. in ceramic engineering from Alfred University and his M.S. and Ph.D. in ceramic engineering from the University of Illinois. He pioneered the development of the materials systems and process technologies that enabled the application of the multilayer ceramic technology to the high-performance packaging applications in IBM's computer systems. While with IBM he also led the advanced materials research and development efforts associated with IBM's storage products. After leaving IBM, Wilcox spent four years as a member of the faculty at the University of Illinois, Department of Materials Science and Engineering. His teaching and research efforts were associated with electronic packaging. In his current position as Director of Advanced Technology, Wilcox manages the Motorola Ceramic Technologies Research Laboratory, which is focused on the development and application of the enabling high-performance ceramic technologies to enhance Motorola's electronic products.

Winder, Stephen M.

Stephen M. Winder was born and educated in the United Kingdom. He gained a Ph.D. in physics of materials (SiAlON ceramics) from Warwick University, and completed a postdoctoral study of glass-ceramic armor for the British Ministry of Defense. He joined British Petroleum Research, London, U.K., in 1985, developed SiC-matrix and Si_3N_4-matrix composite ceramic materials, and invented and patented the thermodynamic-step sintering process. He transferred to BP Research, Cleveland, OH, in 1990 to work with Carborundum's Advanced Ceramics business unit. He transferred to Carborundum Niagara Falls Technical Center in 1993 to lead research projects for the Fused-Cast Refractories, SiC-Refractories, and Fibers business units. Since 1996 he has worked as a technical consultant to Monofrax Refractories, Inc., with major interests in application and development of materials.

Wood, Thomas E.

Thomas E. Wood received a Ph.D. in inorganic chemistry from Purdue University in 1980 and since that time has been working at 3M Company, St. Paul, MN, in the field of sol–gel chemistry and ceramic materials. Wood's research focuses on the chemistry of sol–gel precursors, metal-ion hydrolysis, colloidal science, particle precipitation, and aluminum complex chemistry. Wood holds patents in the areas of ceramic fibers, ceramic beads, transparent ceramics, alumina–chromia abrasive minerals, ceramic bubbles, nucleating agents for α-alumina, and other ceramic processes. As a hobby, Wood studies the history of sol–gel chemistry and ceramic technologies and has cowritten a review on this subject (T. E. Wood and H. Dislich, "An Abbreviated History of Sol–Gel Technology"; in Ceramic Transactions, Vol. 55, *Sol–Gel Science and Technology*. Edited by E. J. A. Pope, S. Sakka, and L. C. Klein. American Ceramic Society, Westerville, OH, 1995).

Yamada, Tetsuo

Tetsuo Yamada is a Senior Researcher at the Ube Research Laboratory at UVE Industries, Ltd., Yamaguchi, Japan. He graduated from the Chemical Institute at Osaka University in 1973 and earned his M.S. in chemistry from Kyoto University in 1975. He received his Ph.D. in engineering from Osaka University in 1983. He has been working on the development and evaluation of new ceramic powders by chemical synthesis routes for engineering applications.

Yasrebi, Mehrdad

Mehrdad Yasrebi received his B.S. in electrical engineering and his M.S. in ceramic engineering from the University of California at Los Angeles in 1981 and 1983, respectively. He is a Senior Research Engineer at PCC Structurals, Inc., a Division of Precision Castparts Corporation, Portland, OR. He holds a Ph.D. in ceramic engineering from the University of Washington, awarded in 1988. Yasrebi's research interests include powder processing of ceramics and metals.

Youngquist, R. J.

R. J. Youngquist received the B.S. and M.S. degrees in electrical engineering from the University of Minnesota. He was a Corporate Scientist for 3M Company. He worked on all aspects of magnetic recording, including tape testing, equipment design, and design of magnetic recording devices. He received the Carlton Society award.

Bibliography for Ceramic Innovations

R.S. Bates, *Scientific Societies in the United States*. The Technology Press, Massachusetts Institute of Technology, New York, and Wiley, New York, 1945.

A. L. Bement Jr. "The Greening of Materials Science and Engineering," *Metall. Trans. A*, **18A** [Mar.] 363–75 (1987).

N. W. Bins, "Twenty-Five Years of the American Ceramic Society," *J. Am. Ceram. Soc.*, **6** [5] 10–20 (1923).

J. E. Burke, "A History of the Development of a Science of Sintering"; pp. 315–33 Ceramics and Civilization, Vol. I, *Ancient Technology to Modern Science*. Edited by W. D. Kingery and E. Lense. American Ceramic Society, Columbus, OH, 1984.

D. Cardwell, *The Norton History of Technology*. W. W. Norton, Worcester, MA, 1994.

L. S. De Camp, *The Ancient Engineers*. Ballantine Books, New York, 1974.

R. W. Douglas and S. Frank, *A History of Glassmaking*. G. T. Foulis, Henly-on-Thames, Oxfordshire, U.K., 1972.

R. B. Gordon and P. M. Malone, *The Texture of Industry—An Archaeological View of the Industrialization of North America*. Oxford University Press, New York, 1994.

B. Hayden, *Archaeology—The Science of Once and Future Things*. W. H. Freeman, New York, 1993.

P. A. Janeway, "ACerS Centennial Celebration—The Formative Years: 1899–1934," *Am. Ceram. Soc. Bull.*, **76** [5] 87–94 (1997).

P. A. Janeway, "Ceramic Pioneers: A. V. Bleininger 1872–1946," *Am. Ceram. Soc. Bull.*, **76** [8] 109–23 (1997).

W. D. Kingery, "An Unseen Revolution: The Birth of High Tech Ceramics"; pp. 293–324 in Ceramics and Civilization, Vol. V, *The Changing Roles of Ceramics in Society: 26,000 B.C. to the Present*. Edited by W. D. Kingery. American Ceramic Society, Westerville, OH, 1990.

C. G. Marvin, "Innovations in Refractories"; presented at the Federation of Materials Societies Conference on Advances in Materials Technology for the Process Industries Needs, 1984. Available from the Refractories Institute, Pittsburgh, PA.

P. McCray and W. D. Kingery (Eds.), Ceramics and Civilization, Vol. VIII, *The Prehistory and History of Glassmaking Technology*. American Ceramic Society, Westerville, OH, 1998.

J. G. Mohr and W. P. Rowe, *Fiber Glass*. Van Nostrand Reinhold, New York, 1978.

R. Newcomb Jr., *Ceramic Whitewares: History, Technology, and Applications*. Pitman Publishing, New York, 1947.

E. Orton Jr., "The American Ceramic Society, Historical Statement of the Origin of The American Ceramic Society," *J. Am. Ceram. Soc.*, **6** [5] 2–10 (1923).

C. W. Pursell Jr. (Ed.), *Readings in Technology and American Life*. Oxford University Press, New York, 1969.

D. W. Readey, "The Response of Ceramic Engineering Education to the Changing Role of Ceramics in Industry and Society"; pp. 343–78 in Ceramics and Civilization, Vol. V, *The Changing Roles of Ceramics in Society: 26,000 B.P. to the Present*. Edited by W. D. Kingery. American Ceramic Society, Westerville, OH, 1990.

S. Saito (Ed.), *Fine Ceramics*. Elsevier, New York, 1985.

S. J. Schneider (Volume Chair), *Ceramics and Glass*, Vol. 4 of Engineered Materials Handbook. ASM International, Metals Park, OH, 1991.

C. J. Singer, E. J. Holmyard, A. R. Hall, and T. I. Williams (Eds.), *A History of Technology, Vol. V, The Late Nineteenth Century c. 1850 to c. 1900*. Oxford University Press, New York, 1958.

J. B. Wachtman, "National Materials Policy—Evolution and Prospects," *J. Res. Manag. Technol.*, **14** [Jan.] 181–90 (1986).

J. B. Wachtman, "Advances in Ceramics"; pp. 141–67 in *The American Ceramic Society —100 Years*, published by The American Ceramic Society in Celebration of its Centennial 1898–1998. American Ceramic Society, Westerville, OH, 1998.

O. J. Whittemore, "Ceramic Patents: A History of Development"; pp. 307–14 in Ceramics and Civilization, Vol. I, *Ancient Technology to Modern Science*. Edited by W. D. Kingery and E. Lense. American Ceramic Society, Columbus, OH, 1984.

T. I. Williams (Ed.), *A History of Technology, Vol. VI, The Twentieth Century c. 1900 to c. 1950 Part I*. Clarendon Press, Oxford University Press, New York, 1978.

M. Wilson, *American Science and Invention—A Pictorial History*. Bonanza Books/Crown Publishers, New York, 1960.

T. E. Wood and H. Dislich, "An Abbreviated History of Sol–Gel Technology"; pp. 3–25 in Ceramic Transactions, Vol. 55, *Sol–Gel Science and Technology*. Edited by E. J. A. Pope, S. Sakka, and L. C. Klein. American Ceramic Society, Westerville, OH, 1995.

Index